AS/A-LEVEL YEAR 2

STUDENT GUIDE

EDEXCEL

Chemistry

Topics 11–15

Equilibrium II
Acid–base equilibria
Energetics II
Redox II
Transition metals

George Facer

PHILIP ALLAN FOR
HODDER
EDUCATION
AN HACHETTE UK COMPANY

Philip Allan, an imprint of Hodder Education, an Hachette UK company, Blenheim Court, George Street, Banbury, Oxfordshire OX16 5BH

Orders

Bookpoint Ltd, 130 Park Drive, Milton Park, Abingdon, Oxfordshire OX14 4SE

tel: 01235 827827

fax: 01235 400401

e-mail: education@bookpoint.co.uk

Lines are open 9.00 a.m.–5.00 p.m., Monday to Saturday, with a 24-hour message answering service. You can also order through the Hodder Education website: www.hoddereducation.co.uk

© George Facer 2016

ISBN 978-1-4718-5841-3

First printed 2016

Impression number 5 4 3 2 1

Year 2020 2019 2018 2017 2016

This guide has been written specifically to support students preparing for the Edexcel A-level Chemistry examinations. The content has been neither approved nor endorsed by AQA and remains the sole responsibility of the author.

Cover photo TTstudio/Fotolia

Typeset by Integra Software Services Pvt. Ltd, Pondicherry, India

Printed in Italy

Hachette UK's policy is to use papers that are natural, renewable and recyclable products and made from wood grown in sustainable forests. The logging and manufacturing processes are expected to conform to the environmental regulations of the country of origin.

Contents

Content Guidance

Questions & Answers

■ Getting the most from this book

Exam-style questions

Commentary on the questions

Tips on what you need to do to gain full marks, indicated by the icon ⓔ

Sample student answers

Practise the questions, then look at the student answers that follow.

Commentary on sample student answers

Find out how many marks each answer would be awarded in the exam and then read the comments (preceded by the icon ⓔ) showing exactly how and where marks are gained or lost.

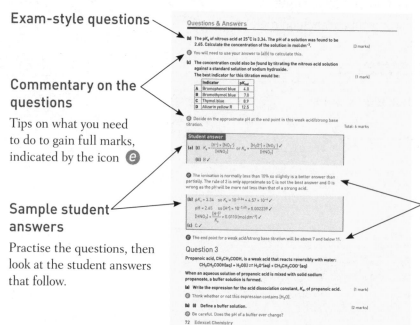

■ About this book

This guide is the third of a series covering the Edexcel specification for A-level chemistry. It offers advice for the effective study of topics 11–15. These five topics (together with topics 1–5 covered in the first student guide of this series, and topics 8 and 10 covered in the second student guide) are examined on A-level paper 1 and as part of synoptic paper 3, which draws on all of the topics 1–19. The guide has two sections:

■ The **Content Guidance** is not intended to be a textbook. It offers guidelines on the main features of the content of topics 11–15, together with advice on making study more productive.

■ The **Questions & Answers** section starts with an introduction that gives advice on approaches and techniques to ensure that you answer the examination questions in the best way you can. It then provides exam-style questions with student answers and comments on how the answers would be marked.

Chemical equations

Equations are quantitative, concise and internationally understood. When you write an equation, check that:

■ you have thought of the *type* of reaction occurring — for example, is it neutralisation, addition or disproportionation?

■ you have written the correct formulae for all the substances

■ your equation balances for the numbers of atoms of each element and for charge

■ you have not written something silly, such as having a strong acid as a product when one of the reactants is an alkali

■ you have included *state symbols* in all thermochemical equations and otherwise if they have been asked for

Diagrams

Diagrams of apparatus should be drawn in section. When you see them, copy them and ask yourself why the apparatus has the features it has. What is the difference between a distillation and a reflux apparatus, for example? When you do practical work, examine each piece of apparatus closely so that you know both its form and its function.

Calculations

Calculations are not normally structured at A-level as they were at AS. Therefore, you need to *plan* the procedure for turning the data into an answer.

■ Set out your calculations in full, making clear what you are calculating at each step. Don't round figures up or down during a calculation. Keep all the numbers on your calculator, or write any intermediate answers to four significant figures.

■ If you have time, check the accuracy of each step by recalculating it. It is so easy to enter a wrong number into your calculator, or to calculate a molar mass incorrectly.

■ Finally, check that you have the correct *units* in your answer and that you have given it to an appropriate number of *significant figures* — if in doubt, give it to three or, for pH calculations, to two decimal places.

Content Guidance

Topic 11 Equilibrium II

Required year 1 knowledge

Dynamic equilibrium

When a mixture reaches equilibrium, the reactions do not stop. The rate of the forward reaction equals the rate of the reverse reaction and there is no further change in the concentrations of any of the substances involved.

Expression for K_c

The equilibrium constant is a measure of the extent to which a reversible reaction takes place at a given temperature. A large value for the equilibrium constant means that a reaction is almost complete. A value greater than 1 means that the reaction proceeds further from left to right than from right to left.

For a **homogeneous** equilibrium, the value of K_c is determined from the *equilibrium* concentrations of all the reactants and products. (Homogeneous means that all the reactants and products are in the same phase, for instance all gases or all in solution.)

For the reaction:

$$xA + yB \rightleftharpoons mC + nD$$

where x, y, m and n are the stoichiometric amounts in the equation, the value of K_c at $T°C$ is given by:

$$K_c = \frac{[C]_{eq}^m [D]_{eq}^n}{[A]_{eq}^x [B]_{eq}^y}$$

The fraction $\frac{[C]^m[D]^n}{[A]^x[B]^y}$ is called the quotient and equals K_c only when the system is at equilibrium.

The relationship for a particular reaction can be shown by measuring the concentrations of the species at equilibrium. The experiment is repeated using several different starting concentrations. For example, in the reaction between hydrogen and iodine:

$$H_2(g) + I_2(g) \rightleftharpoons 2HI(g)$$

Known amounts of hydrogen and iodine are heated to a certain temperature and allowed to reach equilibrium. The system is then analysed to find the amounts of each substance present.

> **Exam tip**
>
> If the quotient is greater than K_c, the system is *not* in equilibrium and the reaction will move to the left until the quotient and K_c are equal.
>
> If the quotient is less K_c, the system is *not* in equilibrium and the reaction will move to the right until the quotient and K_c are equal.

The experiment is repeated using different ratios of hydrogen and iodine and heated to the same temperature. Whatever the starting amounts, the relationship:

$$\frac{[\text{HI}]^2_{eq}}{[\text{H}_2]_{eq}[\text{I}_2]_{eq}}$$

is always equal to the same number — the equilibrium constant, K_c.

Worked example

Consider the reaction:

$$2SO_2(g) + O_2(g) \rightleftharpoons 2SO_3(g) \qquad K_c = 1.7 \times 10^6 \, \text{mol}^{-1} \, \text{dm}^3 \text{ at } 700 \, \text{K}$$

A steel vessel, volume $2 \, \text{dm}^3$, contains $0.2 \, \text{mol} \, SO_3$, $0.04 \, \text{mol} \, SO_2$ and $0.01 \, \text{mol} \, O_2$. Show whether or not this mixture is at equilibrium. If not, indicate which direction the system will move in order to achieve equilibrium at $700 \, \text{K}$.

Answer

$$K_c = \frac{[SO_3]^2_{eq}}{[SO_2]^2_{eq}[O_2]_{eq}} = 1.7 \times 10^6 \, \text{mol}^{-1} \, \text{dm}^3$$

$$[SO_3] = \frac{0.2}{2} = 0.1 \, \text{mol} \, \text{dm}^{-3}$$

$$[SO_2] = \frac{0.04}{2} = 0.02 \, \text{mol} \, \text{dm}^{-3}$$

$$\text{quotient} = \frac{[SO_3]^2}{[SO_2]^2[O_2]} = \frac{0.1^2}{0.02^2} \times 0.005 = 5.0 \times 10^3 \, \text{mol}^{-1} \, \text{dm}^3$$

This value does not equal K_c so the system is *not* at equilibrium.

The quotient (concentration term) is *less* than K_c so its value must increase to achieve equilibrium. This means that the reaction will move to the right, increasing the concentration of SO_3 and decreasing the concentrations of SO_2 and O_2, until the value of the quotient equals 1.7×10^6. The system is then in equilibrium.

Exam tip

Don't forget to divide the number of moles by the volume to get the concentration.

$$[O_2] = \frac{0.01}{2}$$
$$= 0.005 \, \text{mol} \, \text{dm}^{-3}$$

Year 2 equilibrium

Expression for K_p

K_p is the equilibrium constant expressed in terms of partial pressures.

For a homogeneous gas phase reaction

$$xA(g) + yB(g) \rightleftharpoons mC(g) + nD(g)$$

The expression for K_p is:

$$K_p = \frac{p(C)^m \, p(D)^n}{p(A)^x \, p(B)^y}$$

where $p(A)$ is the partial pressure of substance A *at equilibrium*.

The partial pressure of a gas in a mixture of gases is defined as the pressure that the gas would exert if it were alone in the vessel at the same temperature. It is calculated using the formula:

partial pressure = mole fraction × total pressure

where:

mole fraction = $\dfrac{\text{number of moles of that gas}}{\text{total number of moles of gas}}$

The units of K_p can be worked out from the expression for K_p by cancelling the units for the individual partial pressure terms.

Remember that the partial pressure of a gas A in a mixture of gases A, B and C is given by:

$p(A)$ = mole fraction of A × total pressure

$= \dfrac{\text{moles of A}}{(\text{moles of A} + \text{moles of B} + \text{moles of C})} \times \text{total pressure}$

Exam tip

The sum of the partial pressures of all the gases in the system equals the total pressure:

$p(A) + p(B) + p(C) = P(\text{total})$

Worked example

For the equilibrium below, state the expression for K_p and give its units.

$N_2(g) + 3H_2(g) \rightleftharpoons 2NH_3(g)$

Answer

$K_p = \dfrac{p(NH_3)^2}{p(N_2)\,p(H_2)^3}$

The units are:

$\dfrac{\text{atm}^2}{\text{atm}} \times \text{atm}^3 = \dfrac{\text{atm}^2}{\text{atm}^4} = \text{atm}^{-2}$

Exam tip

All pressures in this topic should be measured in atmospheres (atm).

For a heterogeneous reaction involving gases

A heterogeneous system is one in which there are two or more phases, for instance a solid and a gas. Solids do not have a partial pressure and so they are left out of the expression for K_p. Their concentration is fixed and so [solid] does not appear in the expression for K_c either.

Worked example

For the equilibrium below, state the expression for K_p and give its units.

$Fe_2O_3(s) + 3CO(g) \rightleftharpoons 2Fe(s) + 3CO_2(g)$

Answer

$K_p = \dfrac{p(CO_2)^3}{p(CO)^3}$

→

The units are:

$$\frac{atm^3}{atm^3} = \text{no units}$$

Consider the reaction:

$$CaO(s) + CO_2(g) \rightleftharpoons CaCO_3(s)$$

The expression for K_c is:

$$K_c = \frac{1}{[CO_2]} \text{ and has units of } dm^3\,mol^{-1}$$

Calculations involving K_c

In K_c calculations, you are always given the chemical equation plus some data. These data could be concentrations at equilibrium, in which case all you have to do is to put the values into the expression for K_c. However, you are usually given the *initial* amounts and either the amount of one substance at equilibrium or the percentage of a reactant converted. You must use *equilibrium* concentrations, not initial concentrations, in your calculation.

The calculation should be carried out in five steps.

- Step 1 — use the chemical equation to write the expression for K_c.
- Step 2 — construct a table and write in the initial number of moles of each substance.
- Step 3 — write in the table the change in moles of each substance and hence work out the number of moles at equilibrium.
- Step 4 — divide the equilibrium number of moles by the volume to get the concentration in $mol\,dm^{-3}$.
- Step 5 — substitute the *equilibrium* concentrations into the expression for K_c and work out its value. At the same time, work out the units of K_c and include them in your answer.

Worked example

Phosphorus pentachloride (PCl_5) decomposes according to the equation:

$$PCl_5(g) \rightleftharpoons PCl_3(g) + Cl_2(g)$$

1.0 mol of phosphorus pentachloride was added to a 20 dm^3 flask and heated to 180°C. When equilibrium had been established, it was found that 32% of the phosphorus pentachloride had decomposed. Calculate the value of K_c at 180°C.

Answer

Step 1: $K_c = \dfrac{[PCl_3]_{eq}\,[Cl_2]_{eq}}{[PCl_5]_{eq}}$

$\rightarrow$

Exam tip

As 32% of the PCl$_5$ reacted and the initial amount was 1 mol, 0.32 × 1 = 0.32 mol reacted and 0.32 mol of PCl$_3$ and 0.32 mol of Cl$_2$ were formed. Therefore, (1.0 − 0.32) = 0.68 mol of PCl$_5$ remained in the equilibrium mixture.

	PCl$_5$	PCl$_3$	Cl$_2$	Units
Step 2 Initial amount	1.0	0	0	mol
Step 3 Change	−0.32	+0.32	+0.32	mol
Moles at equilibrium	1.0 − 0.32 = 0.68	0.32	0.32	mol
Step 4 Concentration at equilibrium	$\frac{0.68}{20}$ = 0.034	$\frac{0.32}{20}$ = 0.016	$\frac{0.32}{20}$ = 0.016	mol dm^{-3}

Step 5: $K_c = \dfrac{[PCl_3]_{eq}\,[Cl_2]_{eq}}{[PCl_5]_{eq}}$

$= \dfrac{0.016\,\text{mol dm}^{-3} \times 0.016\,\text{mol dm}^{-3}}{0.034\,\text{mol dm}^{-3}} = 0.0075\,\text{mol dm}^{-3}$

The calculation is more complicated if the stoichiometric ratio in the equation is not 1:1. Consider the equilibrium:

$CH_4(g) + 2H_2O(g) \rightleftharpoons CO_2(g) + 4H_2(g)$

For each mole of methane that reacts, 2 mol of steam are required and 1 mol of carbon dioxide and 4 mol of hydrogen are produced.

Worked example

Chlorine can be produced from hydrogen chloride by the following reaction:

$4HCl(g) + O_2(g) \rightleftharpoons 2Cl_2(g) + 2H_2O(g)$

0.40 mol of hydrogen chloride and 0.10 mol of oxygen were placed in a vessel of volume 4.0 dm^3 and allowed to reach equilibrium at 400°C. At equilibrium, only 0.040 mol of hydrogen chloride was present. Calculate the value of K_c.

Answer

$K_c = \dfrac{[Cl_2]^2_{eq}\,[H_2O]^2_{eq}}{[HCl]^4_{eq}\,[O_2]_{eq}}$

	4HCl	O$_2$	2Cl$_2$	2H$_2$O
Initial moles	0.40	0.10	0	0
Change	−0.36	$-\frac{1}{4}$(0.36) = 0.09	$+\frac{1}{2}$(0.36) = +0.18	$+\frac{1}{2}$(0.36) = +0.18
Moles at equilibrium	0.40 − 0.36 = 0.04	0.10 − 0.09 = 0.01	0.18	0.18
Concentration at equilibrium	$\frac{0.04}{4.0}$ = 0.010	$\frac{0.01}{4.0}$ = 0.0025	$\frac{0.18}{4.0}$ = 0.045	$\frac{0.18}{4.0}$ = 0.045

The moles of hydrogen chloride decreased from 0.40 to 0.040, which is a decrease of 0.36 mol.

$K_c = \dfrac{[Cl_2]^2_{eq}\,[H_2O]^2_{eq}}{[HCl]^4_{eq}\,[O_2]_{eq}} = \dfrac{(0.045\,\text{mol dm}^{-3})^2 \times (0.045\,\text{mol dm}^{-3})^2}{(0.010\,\text{mol dm}^{-3})^4 \times 0.0025\,\text{mol dm}^{-3}} = 1.6\,\text{mol}^{-1}\,\text{dm}^3$

Knowledge check 1

The equilibrium constant, K_c, for the reaction:

$H_2(g) + I_2(g) = 2HI(g)$

is 49. **a** Why does K_c for this reaction have no units? **b** Calculate the concentration of HI in an equilibrium mixture containing 0.0040 mol dm^{-3} of I$_2$.

Exam tip

There were 0.04 mol of HCl at equilibrium and the initial amount was 0.40 mol. Therefore, the amount of HCl that reacted was 0.40 − 0.04 = 0.36 mol. The moles of O$_2$ reacted were 0.25 × 0.36 because the ratio of O$_2$ to HCl in the equation is 1:4. The moles of Cl$_2$ (and of H$_2$O) produced were 0.5 × 0.36 because the ratio of moles of Cl$_2$ to HCl in the equation is 2:4.

Calculations involving K_p

In K_p calculations, you are always given the chemical equation plus some data, usually the initial amounts and either the amount of one substance at equilibrium or the percentage of a reactant converted. You will also be given the total pressure. You *must* use equilibrium amounts, *not* initial amounts, in your calculation.

The calculation should be carried out in six steps.
- Step 1 — use the chemical equation to write the expression for K_p.
- Step 2 — construct a table and write in the initial number of moles of each substance.
- Step 3 — write in the table the change in moles of each substance and hence work out the number of moles at equilibrium. Add up the moles to find the *total* number of gas moles.
- Step 4 — divide the number of moles of each substance at equilibrium by the total number of moles to obtain the mole fraction.
- Step 5 — multiply the mole fraction by the *total* pressure to obtain the partial pressure of each gas.
- Step 6 — substitute the partial pressures into the expression for K_p and work out its value. At the same time, work out the units of K_p and include them in your answer.

> **Exam tip**
>
> For a decomposition equilibrium, such as $PCl_5 \rightleftharpoons PCl_3 + Cl_2$, some questions simply state that a percentage of the reactant has reacted. If so, assume that you started with 1 mol and work from that value.

Worked example

Consider the equilibrium:

$$2SO_2(g) + O_2(g) \rightleftharpoons 2SO_3(g)$$

When 1.6 mol of sulfur dioxide was mixed with 1.2 mol of oxygen and allowed to reach equilibrium at 450°C, 90% of the sulfur dioxide reacted. The total pressure in the vessel was 7.0 atm. Calculate the value of K_p at this temperature.

Answer

Step 1: $K_p = \dfrac{p(SO_3)^2}{p(SO_2)^2 \times p(O_2)}$

		$2SO_2$	O_2	$2SO_3$	Total
Step 2	Initial moles	1.6	1.2	0	—
Step 3	Change	−90% of 1.6 = −1.44	−½ × 1.44 = −0.72	+1.44	—
	Moles at equilibrium	1.6 − 1.44 = 0.16	1.2 − 0.72 = 0.48	1.44	2.08
Step 4	Mole fraction	$\dfrac{0.16}{2.08}$ = 0.0769	$\dfrac{0.48}{2.08}$ = 0.231	$\dfrac{1.44}{2.08}$ = 0.692	—
Step 5	Partial pressure atm	0.0769 × 7 = 0.538	0.231 × 7 = 1.62	0.692 × 7 = 4.84	—

Step 6: $K_p = \dfrac{p(SO_3)^2}{p(SO_2)^2 \times p(O_2)} = \dfrac{(4.84\,\text{atm})^2}{(0.538\,\text{atm})^2 \times 1.62\,\text{atm}} = 50\,\text{atm}^{-1}$

> **Exam tip**
>
> 90% of the 1.6 mol SO_2 reacted, so 0.90 × 1.6 mol = 1.44 mol reacted = moles of SO_3 produced.
>
> moles of O_2 that reacted = 0.5 × 1.44 = 0.72
>
> total number of moles = 0.16 + 0.48 + 1.44 = 2.08

A mixture of nitrogen and hydrogen reached equilibrium at 100 atm and 673 K.

$$N_2(g) + 3H_2(g) \rightleftharpoons 2NH_3(g)$$

The partial pressures of nitrogen and hydrogen in the equilibrium mixture were 19 atm and 57 atm. Calculate the value of the equilibrium constant.

Effect of a change in conditions

This is explained by explaining the effect, if any, on the value of the equilibrium constant and on the quotient.

Temperature

A change in temperature is the only factor that alters the value of the equilibrium constant for a given reaction.

Exothermic reactions

The value of K decreases when the temperature is increased. This means that it is now smaller than the quotient, Q, and therefore the position of equilibrium moves to the left until the value of the quotient equals the smaller value of K. The result is that the equilibrium yield is smaller.

Endothermic reactions

The value of K increases when the temperature is decreased. This means that it is now larger than the quotient, Q, and therefore the position of equilibrium moves to the right until the value of the quotient equals the larger value of K. The result is that the equilibrium yield is greater.

Pressure

This will have *no* effect on the value of the equilibrium constant. If there are a different number of gas moles on each side of the chemical equation, the value of the quotient will alter. If it gets bigger, the equilibrium will shift to the left, making the quotient smaller, until it once again equals the unchanged value of K. Likewise if the quotient gets smaller as a result of a change in pressure, the system will move right until the quotient equals the unchanged value of K.

Explain the effect of an increase in pressure on the equilibrium yield for the equilibrium:

$$N_2(g) + 3H_2(g) \rightleftharpoons 2NH_3(g)$$

Answer

$$K_p = \frac{p(NH_3)^2}{p(N_2)\,p(H_2)^3}$$

→

It is easier to explain the effect of an increase in pressure if you assume that the total pressure is doubled.

The value of K_p (or of K_c) does not alter.

If the total pressure is doubled, all the partial pressures are doubled. The numerator (top line) increases by 2^2 and the denominator by 2^4. This makes the quotient smaller and no longer equal to the unchanged value of K. The position moves to the right until the quotient once again equals K.

Exam tip

The rule is that an increase in pressure drives the equilibrium towards the side with fewer gas moles.

Concentration

The value of the equilibrium constant is *not* altered by changes in concentration. However the value of the quotient will change and so the system will no longer be in equilibrium.

Consider the equilibrium reaction:

$$Cr_2O_7{}^{2-}(aq) + H_2O(l) \rightleftharpoons 2CrO_4{}^{2-}(aq) + 2H^+(aq)$$

If concentrated hydrochloric acid is added, $[H^+]$ will increase. This makes the quotient bigger than K_c and so the position will shift to the left until the value of the quotient once again equals the unchanged value of K_c. The colour of the solution will change from the yellow of $CrO_4{}^{2-}$ ions to the orange of $Cr_2O_7{}^{2-}$ ions.

Catalyst

A catalyst has *no* effect on either the value of the equilibrium constant or the value of the quotient. Thus it will not alter the equilibrium yield. Its purpose is for equilibrium to be reached more quickly. In many industrial exothermic processes, this means that the reaction can take place at a reasonable rate at a lower temperature, which results in a higher yield than if it were carried out at a higher temperature. A good example is the Haber process for the manufacture of ammonia.

Distribution of a solute between two immiscible solvents

If a solute such as iodine is shaken between water and a hydrocarbon solvent, an equilibrium is reached where:

$$\frac{[I_2] \text{ in water}}{[I_2] \text{ in the hydrocarbon solvent}} = \text{a constant } K_c$$

This constant is sometimes called the distribution constant or partition constant. Care must be taken to use concentrations not moles, if the volume of the water is different from that of the hydrocarbon solvent.

Summary

Check that you can:
- write expressions for K_c and K_p for homogeneous and heterogeneous equilibria
- calculate the values of K_c and K_p given equilibrium data
- work out the units of K_c and K_p
- explain the effect of changing the temperature on the value of the equilibrium constant and hence on the equilibrium yield.
- understand that the value of the equilibrium constant is neither changed by pressure nor by the addition of a catalyst
- explain the effect of changing the pressure on a gaseous equilibrium

■Topic 12 Acid–base equilibria

Early theories

Acids were first identified by their sour taste. It was then found that acids had a common effect on certain vegetable extracts such as litmus or red cabbage. The next step was the thought that all acids contained oxygen. In fact the name oxygen means 'sharp-maker' in Greek. Later the Dutch chemist Arrhenius suggested that acids produced hydrogen ions in aqueous solution.

$$HA(aq) + H_2O(l) \rightarrow H_3O^+(aq) + A^-(aq)$$

This definition is still used today but only applies to aqueous solutions. Brønsted and Lowry expanded this theory to include a number of similar reactions.

Brønsted–Lowry theory of acidity

Acids

An acid is a proton (H^+ ion) donor; it gives an H^+ ion to a base. Hydrogen chloride is an acid, giving an H^+ ion to a base such as OH^-:

$$HCl + OH^- \rightarrow H_2O + Cl^-$$

An acid also protonates water, which acts as a base:

$$HCl + H_2O \rightarrow H_3O^+ + Cl^-$$

Strong acids protonate water completely, as in the example above. *Weak* acids, such as ethanoic acid, protonate water reversibly:

$$CH_3COOH + H_2O \rightleftharpoons H_3O^+ + CH_3COO^-$$

Bases

A base accepts H^+ ions. To do this, it must have a lone pair of electrons. Ammonia is a base, accepting an H^+ ion from an acid such as hydrogen chloride:

$$NH_3 + HCl \rightarrow NH_4^+ + Cl^-$$

Ammonia is a weak base. It is protonated *reversibly* by water, which acts as an acid:

$$NH_3 + H_2O \rightleftharpoons NH_4^+ + OH^-$$

Exam tip

Acids react with oxides and hydroxides of metals to give a salt and water only. They react with carbonates to give a salt, water and carbon dioxide.

Knowledge check 3

Sulfuric acid is a strong acid in its first ionisation, but a weak acid in its second. Write equations to show this.

Conjugate pairs

Acid-base reactions involve the transfer of protons. The acid loses an H^+ ion and a base accepts it. The species resulting from the loss of an H^+ is called the **conjugate base** of that acid (Table 1). The species resulting from a base gaining an H^+ ion is called the **conjugate acid** of that base.

Acid	Conjugate base
HCl	Cl^-
H_2SO_4	HSO_4^-
H_2O	OH^-
CH_3COOH	CH_3COO^-
NH_4^+	NH_3
HSO_4^-	SO_4^{2-}

Base	Conjugate acid
OH^-	H_2O
H_2O	H_3O^+
NH_3	NH_4^+
$C_2H_5NH_2$	$C_2H_5NH_3^+$
HSO_4^-	H_2SO_4

Table 1

The equation for an acid–base reaction has an acid on the left with its conjugate base on the right, and a base on the left with its conjugate acid on the right (Figure 1). In this type of reaction there are two acid–base conjugate pairs that you must be able to identify.

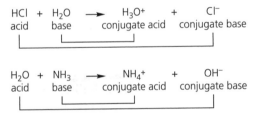

Figure 1 Conjugate pairs

- An acid is linked to its conjugate base by the loss of an H^+ ion.
- A base is linked to its conjugate acid by the gain of an H^+ ion.

Knowledge check 4

Ethanoic acid, CH_3COOH, is a stronger acid than iodic(I) acid, HOI. Complete the following equation:

$$HOI + CH_3COOH \rightarrow$$

Exam tip

Water can act either as a base or as an acid.

Worked example

Concentrated nitric acid reacts with concentrated sulfuric acid according to the equation:

$$HNO_3 + H_2SO_4 \rightarrow H_2NO_3^+ + HSO_4^-$$

Identify the acid–base conjugate pairs in this reaction.

Answer

First pair: acid, H_2SO_4; its conjugate base, HSO_4^-

Second pair: base, HNO_3; its conjugate acid, $H_2NO_3^+$

pH

Definition of pH

The pH of a solution is defined as the negative logarithm to the base 10 of the hydrogen ion concentration in $mol\,dm^{-3}$.

$$pH = -\log[H^+]$$

Thus, if $[H^+] = 1.23 \times 10^{-2}\,mol\,dm^{-3}$, the $pH = -\log[H^+] = -\log(1.23 \times 10^{-2}) = 1.91$.

If $pH = 2.44$, $[H^+] = 10^{-2.44} = 0.00363\,mol\,dm^{-3}$.

Exam tip

Always give pH values to two decimal places. You must know how to convert $[H^+]$ to pH and pH to $[H^+]$ on your calculator.

Definition of pOH

The pOH of a solution is defined as the negative logarithm to the base 10 of the hydroxide ion concentration.

$$pOH = -\log[OH^-]$$

Knowledge check 5

Calculate the pOH of a $0.0246\,mol\,dm^{-3}$ solution of NaOH.

K_w and pK_w

Water partially ionises.

$$H_2O \rightleftharpoons H^+ + OH^-$$

$$K_c = \frac{[H^+][OH^-]}{[H_2O]}$$

The concentration of H_2O is so large that it is effectively constant. Therefore, its value can be incorporated into the expression for the equilibrium constant. Thus, the dissociation constant for water, K_w, is defined as:

$K_w = [H^+][OH^-]$ and at 25°C its value is 1.0×10^{-14}

$pK_w = -\log K_w = 14$

$pK_w = -\log([H^+][OH^-]) = pH + pOH$

At 25°C, $pH + pOH = 14$

Exam tip

This expression is useful for working out the pH of alkaline solutions.

pH of pure water, acidic solutions and alkaline solutions

Neutral solutions

- Pure water is neutral.
  - A *neutral* solution is defined as one in which $[H^+] = [OH^-]$.
  - Therefore, $[H^+] = [OH^-] = \sqrt{K_w} = \sqrt{1.0 \times 10^{-14}} = 1.0 \times 10^{-7}\,mol\,dm^{-3}$ at 25°C.
  - Therefore, the pH of a neutral solution at 25°C $= -\log(1.0 \times 10^{-7}) = 7.00$.
- An *acidic* solution is defined as one in which $[H^+] > [OH^-]$.
  - Therefore, at 25°C, $[H^+] > 1.0 \times 10^{-7}$, which means that $pH < 7$.
- An *alkaline* solution is defined as one where $[H^+] < [OH^-]$.
  - Therefore, at 25°C, $[H^+] < 1.0 \times 10^{-7}$, resulting in a $pH > 7$.
  - If a solution has $[H^+] = 3.0 \times 10^{-9}\,mol\,dm^{-3}$, the $pH = -\log(3.0 \times 10^{-9}) = 8.52$ and it is alkaline.

If the temperature is not 25°C:

- The ionisation of water is endothermic, so an increase in temperature will result in an increase in the value of K_w. At 100°C, the value of $K_w = 5.13 \times 10^{-13}\,mol^2\,dm^{-6}$.
- A neutral solution at 100°C will have $[H^+] = \sqrt{K_w} = \sqrt{5.13 \times 10^{-13}} = 7.16 \times 10^{-7}\,mol\,dm^{-3}$, and so has a pH $= -\log(7.16 \times 10^{-7}) = 6.15$, and *not* 7.

Exam tip

A neutral solution only has a pH of 7 if the temperature is 25°C.

pH scale

- Neutral solutions: $[H^+] = [OH^-]$; pH = 7.00 (at 25°C)
- Acidic solutions: $[H^+] > [OH^-]$; pH < 7
- Alkaline solutions: $[H^+] < [OH^-]$; pH > 7

The pH scale runs from negative numbers to above 14 (Table 2). A small change in hydrogen ion or hydroxide ion concentration causes a large pH change in a solution with an initial pH close to 7. A pH change from 1 to 0 is the result of a change in hydrogen ion concentration from 0.1 to 1 mol dm^{-3} (a change of 0.9 mol dm^{-3}). A pH change from 6 to 5 is the result of a change in hydrogen ion concentration from 0.000 001 to 0.000 01 mol dm^{-3}, an increase of only 0.000 009 mol dm^{-3}.

pH at 25°C	$[H^+]/mol\,dm^{-3}$	Acidity
Negative	> 1	Very strongly acidic
0 to 2	1 to 10^{-2}	Strongly acidic
2 to 5	10^{-2} to 10^{-5}	Weakly acidic
6	10^{-6}	Very weakly acidic
7	10^{-7}	Neutral

pH at 25°C	$[OH^-]/mol\,dm^{-3}$	Alkalinity
7	10^{-7}	Neutral
8	10^{-6}	Very weakly alkaline
9 to 12	10^{-5} to 10^{-2}	Weakly alkaline
12 to 14	10^{-2} to 1	Strongly alkaline
Above 14	> 1	Very strongly alkaline

Table 2 pH, ion concentration and acidity or alkalinity

Calculation of pH of an alkaline solution

The pH of an alkaline solution can be calculated from $[OH^-]$. For example:

$[OH^-] = 2.2 \times 10^{-2}\,mol\,dm^{-3}$

$pOH = -\log[OH^-] = -\log(2.2 \times 10^{-2}) = 1.66$

$pH + pOH = 14$

$pH = 14 - pOH = 14 - 1.66 = 12.34$

An alternative method is:

$[OH^-] = 2.2 \times 10^{-2}\,mol\,dm^{-3}$

$[H^+][OH^-] = 1.0 \times 10^{-14}$

$[H^+] = \dfrac{1.0 \times 10^{-14}}{[OH^-]} = \dfrac{1.0 \times 10^{-14}}{2.2 \times 10^{-2}} = 4.55 \times 10^{-13}$

$pH = -\log[H^+] = -\log(4.55 \times 10^{-13}) = 12.34$

Exam tip

Remember:
pH + pOH = 14 and
$[H^+] = 1.0 \times 10^{-14}/[OH^-]$.

Strong acids and bases

Strong acids

A strong acid, such as hydrochloric acid, is *totally* ionised in aqueous solution:

$$HCl + aq \rightarrow H^+(aq) + Cl^-(aq) \quad \text{or} \quad HCl(aq) + H_2O(l) \rightarrow H_3O^+(aq) + Cl^-(aq)$$

Thus, $[H^+]$ = the initial concentration of the acid.

Hydrochloric acid, hydrobromic acid, HBr, and hydroiodic acid, HI, are all strong acids. Nitric acid, HNO_3, is also a strong acid. Sulfuric acid, H_2SO_4, is strong in its *first* ionisation only:

$$H_2SO_4 + aq \rightarrow H^+(aq) + HSO_4^-(aq)$$

> **Exam tip**
>
> The pH of an acid can be negative if its concentration is more than $1\,mol\,dm^{-3}$. For instance, the pH of a $2.00\,mol\,dm^{-3}$ solution of $HCl = -\log(2.00) = -0.30$.

Worked example

Calculate the pH of the strong monobasic acid HCl of concentration $0.123\,mol\,dm^{-3}$.

Answer

$$pH = -\log[H^+] = -\log(0.123) = 0.91$$

Strong bases

Strong bases are totally ionised in solution. For example:

$$NaOH + aq \rightarrow Na^+(aq) + OH^-(aq)$$

$$Ba(OH)_2 + aq \rightarrow Ba^{2+}(aq) + 2OH^-(aq)$$

For strong bases with one OH^- ion per formula, the OH^- ion concentration is equal to the initial concentration of the base. For bases with two OH^- ions per formula, the OH^- ion concentration is twice the initial concentration of the base.

There are two ways of calculating the pH.

Worked example 1

Calculate the pH of a $2.00\,mol\,dm^{-3}$ solution of sodium hydroxide.

Answer using method 1

$$[OH^-] = 2.00\,mol\,dm^{-3}$$

$$pOH = -\log[OH^-] = -\log(2.00) = -0.30$$

$$pH = 14 - pOH = 14 - (-0.30) = 14.30$$

Answer using method 2

$$[OH^-] = 2.00\,mol\,dm^{-3}$$

$$[H^+] = \frac{1.0 \times 10^{-14}}{[OH^-]} = \frac{1.0 \times 10^{-14}}{2.00} = 5.00 \times 10^{-15}$$

$$pH = -\log[H^+] = -\log(5.00 \times 10^{-15}) = 14.30$$

> **Exam tip**
>
> Remember that pH + pOH = 14

Worked example 2

Calculate the pH of a $0.222 \, mol \, dm^{-3}$ solution of $Ba(OH)_2$.

Answer using method 1

$[OH^-] = 2 \times 0.222 = 0.444 \, mol \, dm^{-3}$

$pOH = -\log[OH^-] = -\log(0.444) = 0.35$

$pH = 14 - pOH = 14 - 0.35 = 13.65$

Answer using method 2

$[OH^-] = 2 \times 0.222 = 0.444 \, mol \, dm^{-3}$

$[H^+] = \dfrac{1.0 \times 10^{-14}}{[OH^-]} = \dfrac{1.0 \times 10^{-14}}{0.444} = 2.25 \times 10^{-14} \, mol \, dm^{-3}$

$pH = -\log[H^+] = -\log(2.25 \times 10^{-14}) = 13.65$

> **Exam tip**
>
> Remember that $[H^+] = K_w/[OH^-]$.

Weak acids

A weak acid is only *slightly* ionised. For dilute solutions of most weak acids, the extent of ionisation is less than 10%, so most of the acid is present as un-ionised molecules.

$CH_3COOH + aq \rightleftharpoons H_3O^+(aq) + CH_3COO^-(aq)$

Acid dissociation constant of a weak acid, K_a

The expression for the acid dissociation constant of a weak acid, and questions depending on it, are easier to follow if the formula of a weak acid is represented by HA.

HA ionises reversibly according to:

$HA \rightleftharpoons H^+ + A^-$ or $HA + H_2O \rightleftharpoons H_3O^+ + A^-$

The acid dissociation constant, K_a, is given by:

$K_a = \dfrac{[H^+][A^-]}{[HA]}$ or $K_a = \dfrac{[H_3O^+][A^-]}{[HA]}$ units: $mol \, dm^{-3}$

The ionisation of one molecule of HA produces one H^+ ion and one A^- ion and so, ignoring the tiny amount of H^+ from the water:

$[H^+] = [A^-]$

The expression for K_a becomes:

$K_a = \dfrac{[H^+]^2}{[HA]}$

$[H^+] = \sqrt{K_a[HA]} = \sqrt{K_a} \times \text{initial concentration of weak acid}$

As the acid is only slightly ionised, you can make the approximation:

$[HA]$ = the initial concentration of the weak acid

Therefore, the pH of a weak acid can be calculated from its concentration and its acid dissociation constant.

> **Exam tip**
>
> You can use H^+, $H^+(aq)$ or H_3O^+ in the chemical equation and in the expression for K_a but do *not* include $[H_2O]$ in the expression for K_a.

> **Exam tip**
>
> Sometimes you are given pK_a not K_a. The pK_a of a weak acid is defined as $-\log K_a$, so $K_a = 10^{-pK_a}$.

Worked example

Calculate the pH of a $0.123\,mol\,dm^{-3}$ solution of ethanoic acid.

$K_a = 1.74 \times 10^{-5}\,mol\,dm^{-3}$

Answer

$CH_3COOH \rightleftharpoons H^+ + CH_3COO^-$

$K_a = \dfrac{[H^+][CH_3COO^-]}{[CH_3COOH]} = \dfrac{[H^+]^2}{[CH_3COOH]} = 1.74 \times 10^{-5}\,mol\,dm^{-3}$

$[H^+] = \sqrt{K_a[CH_3COOH]} = \sqrt{1.74 \times 10^{-5} \times 0.123} = 0.00146\,mol\,dm^{-3}$

$pH = -\log[H^+] = -\log(0.00146) = 2.83$

You must understand the two assumptions made here:

1 $[CH_3COOH]$ in solution = [acid] originally (ignoring the small amount ionised)
2 $[H^+] = [CH_3COO^-]$ (ignoring the small amount of H^+ produced by the ionisation of water)

If you are given the mass of ethanoic acid and the volume of solution, you must first calculate the concentration of the solution in $mol\,dm^{-3}$:

$$\text{concentration} = \text{moles} \div \text{volume in dm}^3 = \frac{\text{mass of acid}}{\text{molar mass of acid}} \div \frac{\text{volume in cm}^3}{1000}$$

Effect of dilution on pH

Worked example 1

Calculate the pH of a $0.123\,mol\,dm^{-3}$ solution of the strong acid HCl, a $0.0123\,mol\,dm^{-3}$ solution and a $0.00123\,mol\,dm^{-3}$ solution.

Answer

For the $0.123\,mol\,dm^{-3}$ solution: $pH = -\log(0.123) = 0.91$

For the $0.0123\,mol\,dm^{-3}$ solution: $pH = -\log(0.0123) = 1.91$

For the $0.00123\,mol\,dm^{-3}$ solution, $pH = -\log(0.00123) = 2.91$

Worked example 2

The pH of a $0.123\,mol\,dm^{-3}$ solution of the weak acid ethanoic acid, $K_a = 1.74 \times 10^{-5}\,mol\,dm^{-3}$, is 2.83 (see the worked example in the previous section). Calculate the pH of a $0.0123\,mol\,dm^{-3}$ solution of ethanoic acid.

Answer

$[H^+]^2 = K_a \times 0.0123 = 1.74 \times 10^{-5} \times 0.0123$

$[H^+] = \sqrt{1.74 \times 10^{-5} \times 0.0123} = 4.63 \times 10^{-4}$

$pH = -\log(4.63 \times 10^{-4}) = 3.33$, which is an increase of half a pH unit from the pH of the solution that is ten times more concentrated.

Exam tip

If you are given pK_a of the acid an extra step has to be made in the calculation. For example: pK_a for ethanoic acid = 4.76, so K_a for ethanoic acid = $10^{-4.76} = 1.74 \times 10^{-5}\,mol\,dm^{-3}$.

Exam tip

When a strong acid is diluted by a factor of 10, its pH increases by 1 unit. When a weak acid is diluted by a factor of 10 its pH increases by half a unit. However, when a buffer solution is diluted by a factor of 10 its pH does not alter.

Calculation of K_a from pH

This is a similar calculation to the previous one. If you are given the pH and the concentration of a solution, K_a can be calculated.

Core practical 9

Calculation of K_a of a weak acid

The method involves weighing out a known mass of the acid, making up the solution to a known volume and measuring its pH.

To find K_a of ethanoic acid:

1 Weigh a weighing bottle empty and then containing about 1.6 g of ethanoic acid.

2 Using a funnel pour the acid into a 250 cm^3 standard flask and wash the contents of the weighing bottle and funnel with distilled water into the flask.

3 Make up to 250 cm^3 and thoroughly shake the flask.

4 Check a pH meter using a buffer solution of known pH.

5 Pour some of the ethanoic acid solution into a beaker and measure its pH.

Readings
mass of weighing bottle empty = 22.17 g
mass of bottle + ethanoic acid = 23.76 g
mass of acid = 1.59 g
pH = 2.87

Calculation

$$K_a = \frac{[H^+][CH_3COO^-]}{[CH_3COOH]}$$

$[H^+] = 10^{-pH} = 10^{-2.87} = 0.00135 = [CH_3COO^-]$

Molar mass $CH_3COOH = 60\,g\,mol^{-1}$

Amount $CH_3COOH = \dfrac{1.59\,g}{60\,g\,mol^{-1}} = 0.0265\,mol$

$[CH_3COOH]_{initial} = \dfrac{0.0265}{0.250} = 0.106\,mol\,dm^{-3}$

$[CH_3COOH]_{equilibrium} = 0.106 - 0.00135 = 0.1047\,mol\,dm^{-3}$

$$K_a = \frac{(0.00135)^2}{0.1047} = 1.74 \times 10^{-5}\,mol\,dm^{-3}$$

As 0.00135 mol of H^+ ions are produced from 0.106 mol of CH_3COOH, the amount of CH_3COOH at equilibrium equals $(0.106 - 0.00135)$.

If $[CH_3COOH] = 0.106\,mol\,dm^{-3}$ had been used, the value of K_a would have been calculated as $1.72 \times 10^{-5}\,mol\,dm^{-3}$ — a less accurate answer that would also have scored full marks.

pH of sulfuric acid

Sulfuric acid is a strong acid from its first ionisation, but a weak acid from its second ionisation:

$$H_2SO_4 + aq \rightarrow H^+ + HSO_4^-$$

$$HSO_4^- + aq \rightleftharpoons H^+ + SO_4^{2-} \qquad K_2 = 0.010\,mol\,dm^{-3}$$

$$K_2 = \frac{[H^+]\,[SO_4^{2-}]}{[HSO_4^-]}$$

Consider a solution of $0.100\,mol\,dm^{-3}$ sulfuric acid:

$[H^+]$ from the first ionisation = $0.100\,mol\,dm^{-3}$

Let $[H^+]$ from second ionisation = z

The second ionisation is driven to the left (suppressed) by the high concentration of H^+ from the first ionisation:

$$K_2 = 0.010 = \frac{(0.100 + z)z}{(0.100 - z)}$$

This can be solved using quadratic equations or by making the approximations $(0.100 - z)$ and $(0.100 + z) \approx 0.100$. Then:

$$K_2 = z = 0.010$$

$$pH = -\log(0.100 + 0.010) = 0.96$$

Thus the pH of a $0.1\,mol\,dm^{-3}$ solution of sulfuric acid is only just below 1, as $0.1\,mol\,dm^{-3}$ of H^+ ions is produced from the first ionisation and hardly any from the second ionisation. If it were assumed that the second ionisation was complete, the pH would be $-\log(0.200) = 0.70$.

pH of various solutions

The pHs of strong and weak acids, strong and weak bases and salts, all at a concentration of $0.1\,mol\,dm^{-3}$, are as shown in Table 3.

Type	Example	pH
Strong acid	Hydrochloric acid (HCl)	1
Weak acid	Ethanoic acid (CH_3COOH)	Approx. 3
Strong base	Sodium hydroxide (NaOH)	13
Weak base	Ammonia (NH_3)	Approx. 11
Salt of a strong acid and a strong base	Sodium chloride (NaCl)	7
Salt of a weak acid and a strong base	Sodium ethanoate (CH_3COONa)	Approx. 9
Salt of a strong acid and a weak base	Ammonium chloride (NH_4Cl)	Approx. 5

Table 3

The rule of two

Note that:

- a weak acid has a pH about 2 units more (less acidic) than a strong acid
- a weak base has a pH about 2 units less (less alkaline) than a strong base
- the salt of a weak acid has a pH about 2 units more than the salt of a strong acid
- the salt of a weak base has a pH about 2 units less than the salt of a strong base

Acid–base titrations

The usual method is to add a standard solution of base from a burette to a known volume of acid, in the presence of a suitable indicator. The addition of base is normally stopped when the 'end point' has been reached.

- The pH at the end point is 7 only when a strong base is titrated with a strong acid.
- The end point is on the acidic side of neutral if a weak base, such as ammonia solution, is titrated with a strong acid and above 7 if a strong base is titrated with a weak acid, such as ethanoic acid (Table 4).

Titration	pH at end point	Common indicators
Strong acid–strong base	7	Methyl orange, methyl red or phenolphthalein
Strong acid–weak base	5–6	Methyl orange or methyl red
Weak acid–strong base	8–9	Phenolphthalein

Table 4 Acid–base titrations

- Acid–base titrations can also be carried out by adding a standard solution of acid from a burette to a known volume of base.

Titration curves

These need to be drawn carefully. Follow these steps:

- Decide whether the acid and base are strong or weak.
- Draw the axes. The y-axis (vertical) should be labelled 'pH', with a linear scale of 0 to 14, and the x-axis labelled 'volume of solution being added' (normally the base). Some questions ask for the titration curve when acid is added to a base, in which case the x-axis represents the volume of acid.
- Calculate the volume that has to be added to reach the end point. In most questions, the concentrations of acid and base are the same; therefore, the volume of base at the end point will be the same as the initial volume of acid.
- Estimate the pH:
  - at the start
  - at the end point
  - at the midpoint of the vertical range of the curve
  - after excess base has been added (final pH)
- Draw a smooth curve with a vertical line at the end point volume.

Exam tip

The end point is reached when enough base has been added to react completely with the acid.

Knowledge check 7

State the colour change observed when sodium hydroxide is added to dilute hydrochloric acid until the end point is reached using **a** methyl orange and **b** phenolphthalein as the indicator.

Table 5 shows pH values during typical titrations.

Type of titration	Starting pH	End point pH	Vertical range of pH	Final pH
Strong base added to strong acid	1	7	3–11	Just <13
Strong base added to weak acid	3	9	7–11	Just <13
Weak base added to strong acid	1	5	3–7	Just <11
Strong acid added to strong base	13	7	11–3	Just >1
Strong acid added to weak base	11	5	7–3	Just >1
Weak acid added to strong base	13	9	11–7	Just >3

Table 5 Type of titration and range of pH values

The graphs in Figure 2 are examples of titration curves.

Strong acid–strong base

The left-hand graph in Figure 2 shows the change in pH when $40\,cm^3$ of base is added to $20\,cm^3$ of acid, both having the same concentration. The right-hand graph is for the addition of $40\,cm^3$ of acid to $20\,cm^3$ of base. The end point is $20\,cm^3$. For example, hydrochloric acid and sodium hydroxide:

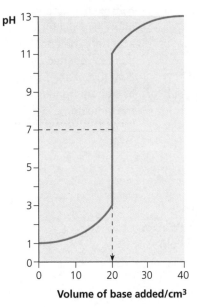
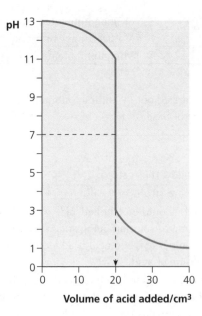

Figure 2

> **Exam tip**
>
> Do not learn this table. Use the rule of 2 on the previous page to estimate these pH values.

Strong base–weak acid

Figure 3 shows sodium hydroxide added to ethanoic acid as an example.

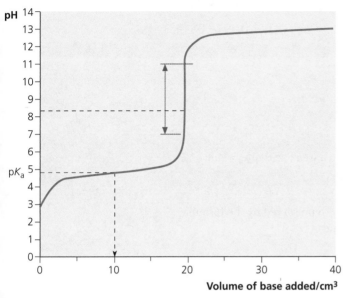

Figure 3

> ### Exam tip
>
> The graph should start just below pH = 3, rise rapidly to a pH of about 4.5 and then rise slowly until just before the end point where is rises vertically and then flattens off to a pH of just less than 13.

Strong acid–weak base

Figure 4 shows an example of hydrochloric acid added to ammonia.

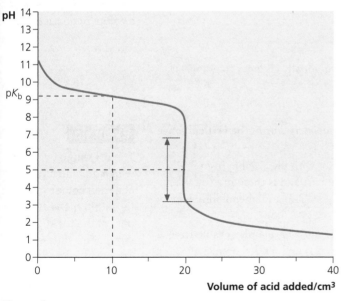

Figure 4

Calculation of pK_a from titration curves

When a strong base is added to a solution of a weak acid, the value of pK_a can be found by reading the pH at the point when *half* the acid has been neutralised by the alkali (half way to the end-point volume).

At this point:

$$[HA]_{halfway} = [A^-]_{halfway}$$

$$K_a = \frac{[H^+][A^-]}{[HA]} = \frac{[H^+]_{halfway}[A^-]_{halfway}}{[HA]_{halfway}} = [H^+]_{halfway}$$

$$pK_a = (pH)_{halfway}$$

In the titration curve for a strong base being added to a weak acid, $pK_a = 4.8$.

Indicators

Indicators are weak acids in which the acid molecule, represented by the formula HInd, is a different colour from its conjugate base, Ind$^-$:

$$HInd \rightleftharpoons H^+ + Ind^-$$

An example is methyl orange. The acid, HInd, is red and its conjugate base, Ind$^-$, is yellow.

$$K_{ind} = \frac{[H^+][Ind^-]}{[HInd]}$$

At the end-point colour, the acid and its conjugate base are in equal proportions. Therefore:

$$[HInd] = [Ind^-]$$

$$K_{ind} = \frac{[H^+][Ind^-]}{[HInd]} = [H^+]_{end\ point}$$

$$pK_{ind} = (pH)_{end\ point}$$

Methyl orange, $pK_{ind} = 3.7$, is red below a pH of 2.7, gradually changes to orange at pH 3.7 and then to yellow at pH 4.7.

Choice of indicator

- For all titrations, the value of $pK_{ind} \pm 1$ must lie *completely* within the vertical range of the titration curve.
- For a strong acid–strong base titration, any indicator with pK_{ind} value from 3.5 to 10.5 has its range completely within the vertical part and is therefore suitable. Thus, methyl red ($pK_{ind} = 5.1$), methyl orange ($pK_{ind} = 3.7$) and phenolphthalein ($pK_{ind} = 9.3$) are all suitable.
- For a strong acid–weak base titration, the indicator must have a pK_{ind} value from 3.5 to 6.5 to be suitable. Thus, methyl red ($pK_{ind} = 5.1$) and methyl orange ($pK_{ind} = 3.7$) are both suitable.
- For a weak acid–strong base titration, the indicator must have a pK_{ind} value between pH 6.5 and 10.5. Phenolphthalein ($pK_{ind} = 9.3$) is suitable.

Exam tip

Most indicators change colour completely over a range of about ± 1 pH unit.

Exam tip

The pH range over which an indicator changes colour must lie within the vertical part of the titration curve. The values are given in the *Data Booklet*.

Buffer solutions

Definition

A buffer solution *resists* a change in pH when *small* amounts of acid or base are added. It is made up of a weak acid and its conjugate base (the salt of the weak acid), or a weak base and its conjugate acid. Examples are:

- ethanoic acid (CH_3COOH) plus sodium ethanoate ($CH_3COO^-Na^+$), which is an acid buffer with a pH less than 7
- ammonia solution (NH_3) plus ammonium chloride ($NH_4^+Cl^-$), which is an alkaline buffer with a pH greater than 7

A buffer solution is at its most efficient when:

- for an acidic buffer, the concentration of the weak acid equals the concentration of its conjugate base
- for an alkaline buffer, the concentration of the weak base equals the concentration of its conjugate acid

> **Exam tip**
>
> Do not state that a buffer solution has a constant pH because the pH does alter slightly even when small amounts of H^+ or OH^- ions are added.

Calculation of pH of a buffer solution

Calculation of the pH of a buffer solution relies on the relationship between K_a and the concentrations of the weak acid, HA, and its salt.

$$K_a = \frac{[H^+][A^-]}{[HA]} = \frac{[H^+][\text{salt of weak acid}]}{[\text{weak acid}]}$$

$$[H^+] = \frac{[K_a][\text{weak acid}]}{[\text{salt of weak acid}]}$$

The pH is then calculated as $-\log[H^+]$.

> **Exam tip**
>
> Note that calculation of the pH of weak acids *and* buffers uses the same expression for K_a. The difference is that for a weak acid $[A^-] = [H^+]$ whereas in a buffer $[A^-] = [\text{salt}]$.

Worked example 1

Calculate the pH of a buffer solution made by adding 6.15 g of sodium ethanoate, CH_3COONa (molar mass $82\,\text{g mol}^{-1}$), to $50.0\,\text{cm}^3$ of a $1.00\,\text{mol dm}^{-3}$ solution of ethanoic acid ($K_a = 1.75 \times 10^{-5}\,\text{mol dm}^{-3}$).

Answer

$$K_a = \frac{[H^+][CH_3COO^-]}{[CH_3COOH]}$$

$$[H^+] = \frac{K_a[CH_3COOH]}{[CH_3COO^-]}$$

$$[CH_3COOH] = 1.00\,\text{mol dm}^{-3}$$

$$\text{amount of sodium ethanoate} = \frac{6.15\,\text{g}}{82\,\text{g mol}^{-1}} = 0.075\,\text{mol}$$

$$[CH_3COO^-] = [CH_3COONa] = \frac{0.075\,\text{mol}}{0.050\,\text{dm}^3} = 1.50\,\text{mol dm}^{-3}$$

$$[H^+] = \frac{1.75 \times 10^{-5}\,\text{mol dm}^{-3} \times 1.00\,\text{mol dm}^{-3}}{1.50\,\text{mol dm}^{-3}} = 1.17 \times 10^{-5}\,\text{mol dm}^{-3}$$

$$\text{pH} = -\log(1.17 \times 10^{-5}) = 4.93$$

Worked example 2

Calculate the pH of a buffer solution made by mixing $100\,cm^3$ of a $0.200\,mol\,dm^{-3}$ solution of ethanoic acid, $K_a = 1.75 \times 10^{-5}\,mol\,dm^{-3}$, with $100\,cm^3$ of a $0.300\,mol\,dm^{-3}$ solution of sodium ethanoate.

Answer

$[CH_3COOH]$ in buffer = ½ × 0.200 = $0.100\,mol\,dm^{-3}$

$[CH_3COO^-]$ in buffer = ½ × 0.300 = $0.150\,mol\,dm^{-3}$

$$[H^+] = \frac{K_a\,[\text{acid}]}{[\text{salt}]} = \frac{1.75 \times 10^{-5} \times 0.100}{0.150} = 1.17 \times 10^{-5}\,mol\,dm^{-3}$$

$pH = -\log(1.17 \times 10^{-5}) = 4.93$

Note that the pH values of the buffer solutions in the two worked examples are the same. The concentrations of acid *and* salt in example 2 are ten times less than the concentrations in example 1. As the ratio [acid]:[salt] is the same, the pH is the same.

A more difficult calculation is when a known volume of strong base solution is added to an *excess* of weak acid solution. All the base reacts with some of the weak acid. Thus a mixture of excess weak acid and its salt is formed, which is a buffer solution.

> **Exam tip**
>
> Diluting a buffer solution does *not* alter its pH.

Worked example 3

Calculate the pH of a buffer solution made by adding $50\,cm^3$ of $1.00\,mol\,dm^{-3}$ solution of sodium hydroxide to $80\,cm^3$ of $1.00\,mol\,dm^{-3}$ solution of propanoic acid (CH_3CH_2COOH), $K_a = 1.35 \times 10^{-5}\,mol\,dm^{-3}$.

Answer

amount of sodium hydroxide added = $1.00\,mol\,dm^{-3} \times 0.050\,dm^3$

$= 0.050\,mol$

$=$ amount of sodium propanoate produced

amount of propanoic acid initially = $1.00\,mol\,dm^{-3} \times 0.080\,dm^3 = 0.080\,mol$

amount of propanoic acid *left* after reaction with sodium hydroxide $= 0.080 - 0.050 = 0.030\,mol$

$[CH_3CH_2COOH] = \dfrac{0.030\,mol}{0.130\,dm^3} = 0.231\,mol\,dm^{-3}$

$[CH_3CH_2COO^-] = [CH_3CH_2COONa] = \dfrac{0.050\,mol}{0.130\,dm^3} = 0.385\,mol\,dm^{-3}$

$K_a = \dfrac{[H^+][CH_3CH_2COO^-]}{[CH_3CH_2COOH]}$

$[H^+] = \dfrac{K_a[CH_3CH_2COOH]}{[CH_3CH_2COO^-]}$

$= \dfrac{1.35 \times 10^{-5}\,mol\,dm^{-3} \times 0.231\,mol\,dm^{-3}}{0.385\,mol\,dm^{-3}} = 8.10 \times 10^{-6}\,mol\,dm^{-3}$

$pH = -\log(8.10 \times 10^{-6}) = 5.09$

> **Exam tip**
>
> Note that propanoic acid and sodium hydroxide react in a 1:1 ratio to produce sodium propanoate and water, and also that the total volume of the solution is 50 + 80 = $130\,cm^3$ = $0.130\,dm^3$.

Summary of pH value calculations for solutions of weak acids and buffers

Both are based on the expression for K_a:

$$K_a = \frac{[H^+][A^-]}{[HA]} \quad \text{so} \quad [H^+] = \frac{K_a[HA]}{[A^-]}$$

In *both* solutions: $\qquad\qquad$ [HA] = [weak acid]

In a solution of a *weak acid*: $\quad$ $[H^+] = [A^-]$

In a *buffer* solution: $\qquad\quad$ $[A^-]$ = [salt]

Mode of action of a buffer solution

Acid buffer

The ethanoic acid–sodium ethanoate buffer system is used as an example. The weak acid is *partially* ionised:

$$CH_3COOH \rightleftharpoons H^+ + CH_3COO^-$$

Its salt, which contains the conjugate base, is *totally* ionised:

$$CH_3COONa \rightarrow CH_3COO^- + Na^+$$

The CH_3COO^- ions produced suppress the ionisation of the weak acid. Therefore:

$[CH_3COOH]$ = [weak acid]

$[CH_3COO^-]$ = [salt of weak acid]

If a *small* quantity of H^+ ions is added, almost all react with the large reservoir of CH_3COO^- ions from the salt:

$$H^+ + CH_3COO^- \rightleftharpoons CH_3COOH$$

The values of $[CH_3COO^-]$ and $[CH_3COOH]$ hardly alter, as they are both large *relative* to the *small* amount of H^+ added.

$$[H^+] = \frac{K_a[\text{acid}]}{[\text{salt}]} = \frac{K_a[CH_3COOH]}{[CH_3COO^-]}$$

If neither $[CH_3COOH]$ nor $[CH_3COO^-]$ alter significantly, the value of $[H^+]$ and hence the pH will also not alter significantly.

If a *small* amount of OH^- ions is added to the buffer solution, almost all of them react with the large reservoir of CH_3COOH molecules from the acid:

$$OH^- + CH_3COOH \rightarrow H_2O + CH_3COO^-$$

Again, neither $[CH_3COOH]$ nor $[CH_3COO^-]$ alters significantly as both are large *relative* to the *small* amount of OH^- added, so the pH hardly alters.

The crucial points that must be made when explaining the mode of action of a buffer are that:

■ almost all the added H^+ ions are removed by reaction with the conjugate base — include equation

> **Knowledge check 8**
>
> Explain why a solution of a weak acid does not buffer small additions of OH^- ions.

- almost all the added OH^- ions are removed by reaction with the weak acid — include equation
- the ratio [weak acid]/[salt] does not change significantly because both quantities are large *relative* to the small amounts of H^+ or OH^- added

> **Exam tip**
>
> Remember that a weak acid ionises *reversibly* and a salt ionises *totally*.

Summary

After studying this topic, you should be able to:
- define an acid, a base, a strong acid and a weak acid
- identify Brønsted–Lowry acid–base pairs
- write the expression for K_a of a weak acid and calculate K_a from pK_a
- calculate the pH of solutions of strong acids, weak acids and strong bases
- state the effect of diluting solutions of strong acids, weak acids and buffers
- draw titration curves and use them to calculate K_a and to identify suitable indicators
- define a buffer solution and calculate its pH
- explain the mode of action of a buffer solution

∎ Topic 13 Energetics II

Topic 13A Lattice energy

Definitions

Lattice energy, LE, is the energy change for the formation of 1 mole of ionic solid from its isolated gaseous ions.

For magnesium chloride, it is the heat change for:

$$Mg^{2+}(g) + 2Cl^-(g) \rightarrow MgCl_2(s)$$

Standard enthalpy of atomisation, $\Delta_{at}H^\ominus$, is the enthalpy change for the formation of 1 mole of gaseous atoms from an element in its standard state at 100 kPa pressure and a specified temperature.

Note carefully that the definition of $\Delta_{at}H$ refers to the *formation of 1 mole of gaseous atoms*. Therefore, in the case of chlorine, the change measured is for the process:

$$\tfrac{1}{2}Cl_2(g) \rightarrow Cl(g)$$

The **first electron affinity** of an atom is the energy change for addition of one electron to each of a mole of atoms in the gas phase:

$$X(g) + e^- \rightarrow X^-(g)$$

This is exothermic, since the electron is attracted to the nuclear charge.

The **second electron affinity** is the energy change per mole for:

$$X^-(g) + e^- \rightarrow X^{2-}(g)$$

This is endothermic, since the incoming electron is repelled by the negative charge on the ion.

> **Exam tip**
>
> Lattice energies are always exothermic ($\Delta_{latt}H$ is negative).

Born–Haber cycles

The formation of an ionic crystal from its elements can be broken down into a series of stages for which the energy changes can be measured. This enables calculation of the **lattice energy**, which, since it is a measure of the attraction between the ions, is a measure of how strongly the crystal is bonded.

The lattice energy cannot be measured directly; if an ionic crystal is heated it might vaporise or it might decompose before it does so.

However, there are other processes for which the energy change can be measured and which, when put together in a Born–Haber cycle, enable a value for the lattice energy to be calculated. Such values are called **experimental (or actual) values**. The lattice energy can also be calculated from the geometry of the crystal and the use of Coulomb's law, which gives a value for the **theoretical lattice energy**, or the lattice energy for a *purely* ionic model. Experimental lattice energy and theoretical lattice energy are seldom the same, because most ionic crystals also have some covalent bonding.

The Born–Haber cycle for sodium chloride is shown in Figure 5.

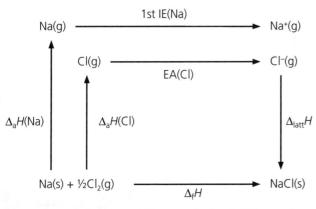

$$\Delta_f H = \Delta_a H(\text{Na}) + \Delta_a H(\text{Cl}) + \text{1st IE(Na)} + \text{EA(Cl)} + \Delta_{latt} H$$

Figure 5

The lattice energy can then be calculated. For sodium chloride, see Table 6.

$\Delta_a H(\text{Na})$	$\Delta_a H(\text{Cl})$	1st IE(Na)	1st EA(Cl)	$\Delta_f H(\text{NaCl})$
+107	+122	+496	−349	−411

Table 6

$$\Delta_{latt} H = -411 - (107 + 122 + 496 - 349) = -787\,\text{kJ mol}^{-1}$$

Polarisation of ions: covalence in ionic lattices

Ionic and covalent bonds are extremes. Ionic bonds involve complete electron transfer and covalent bonds involve equal sharing of electrons. In practice, bonding in most compounds is intermediate between these two forms, with one type being predominant.

Exam tip

Remember that $\Delta_a H$ of chlorine is the enthalpy change for $\frac{1}{2}\text{Cl}_2(g) \rightarrow \text{Cl}(g)$.

The attraction that a bonded atom has for electrons is called its **electronegativity**. There are several electronegativity scales, the commonest being the Pauling scale. Fluorine is the most electronegative element with a value of 4, caesium the least at 0.7.

- Atoms with the same electronegativity bond covalently, with equal sharing of electrons.
- Atoms with different electronegativities form polar covalent bonds if the difference is not too large — up to about 1.5 — and ionic bonds if the difference is greater.

Cations are able to distort the electron clouds of anions, which generally have a larger radius. This leads to a degree of electron sharing and hence some covalence in most ionic compounds. Cations with a small radius and a high charge (2+ or 3+) have a high charge density. They can polarise anions, and therefore have a high **polarising power**. Anions with a large radius do not hold on to their outer electrons very tightly, so they can be easily distorted — such ions are **polarisable**. Ions of charge 2– are more polarisable than those that are 1–.

The effect of polarisation is that the lattice energy of these compounds is different from that predicted on the basis of a purely ionic model; the lattice energy is more exothermic, so the lattice is stronger. The smaller the cation associated with a given anion, the greater is the degree of covalency (Table 7). This is because smaller cations have higher charge densities.

> **Exam tip**
>
> The bigger the difference between the experimental and theoretical lattice energies, the greater is the extent of covalency.

Compound	$\Delta_{latt}H$/kJ mol^{-1} (ionic model)	$\Delta_{latt}H$/kJ mol^{-1} (Born–Haber)	Difference/kJ mol^{-1}
$MgCl_2$	−2326	−2526	200
$CaCl_2$	−2223	−2258	35
$SrCl_2$	−2127	−2156	29
$BaCl_2$	−2056	−2033	23

Table 7 The smaller the cation the greater the degree of covalency

The larger the anion associated with a given cation, the greater is the degree of covalency (Table 8). This is because the larger anions are more polarisable.

Compound	$\Delta_{latt}H$/kJ mol^{-1} (ionic model)	$\Delta_{latt}H$/kJ mol^{-1} (Born–Haber)	Difference/kJ mol^{-1}
MgF_2	−2913	−2957	44
$MgCl_2$	−2326	−2526	200
$MgBr_2$	−2097	−2440	343
MgI_2	−1944	−2327	383

Table 8 The larger the anion the greater the degree of covalency

> **Knowledge check 9**
>
> Put the following in order of increasing difference between experimental and theoretical lattice energies: NaCl, $MgBr_2$, $MgCl_2$.

Solubility

The solubility of a substance in a given solvent depends on the enthalpy (heat energy) and entropy changes of the dissolving process. Entropy is considered on page 34, so solubility is discussed here only in terms of enthalpy, but you should remember that the full story is more complex.

In general, a substance dissolves if the forces between the solute and the solvent are of similar or greater strength than those that are broken both between the solute molecules and between the solvent molecules. Thus, for a solute A and a solvent B, the A–B forces must be similar in strength to, or stronger than, both the A–A forces and the B–B forces.

Ionic compounds

When an ionic compound dissolves in water, the ionic lattice has to be broken up (endothermic process) and the resulting ions are then hydrated (exothermic process).

Definitions

The **enthalpy of solution**, $\Delta_{sol}H$, of an ionic solid is the enthalpy change when 1 mole dissolves in water forming ions infinitely apart.

The **hydration enthalpy**, $\Delta_{hyd}H$, of an ion is the enthalpy change when 1 mole of the gaseous ions dissolves in water until there is no further heat change.

Ions with a small radius and large charge have a more exothermic hydration enthalpy than bigger, less charged ions.

The enthalpy of solution, the lattice energy and the hydration energies of the ions are connected using a Hess's law cycle (Figure 6).

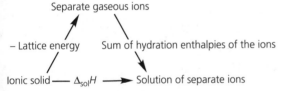

Figure 6

If the energy required to break up the lattice is recovered by the hydration of the ions, the compound is likely to be soluble.

$$\Delta_{sol}H = -\Delta_{latt}H + \Delta_{hyd}H(\text{cation}) + \Delta_{hyd}H(\text{anion})$$

$\Delta_{sol}H$ must be exothermic or only slightly endothermic. (Remember that this applies only where the entropy changes are less important than the enthalpy changes.)

Factors determining lattice energy

Ionic charge

The greater the charge, the stronger is the force between the positive and the negative ions. Therefore, the lattice energy is more exothermic. For example, the lattice energy of magnesium fluoride (MgF_2), which contains the Mg^{2+} ion, is much more exothermic than that of sodium fluoride (NaF), which contains the Na^+ ion.

Sum of the ionic radii

The smaller the sum of the ionic radii, the stronger is the force between the positive and the negative ions. Therefore, the lattice energy is more exothermic. For example:

- the lattice energy of sodium fluoride is more exothermic than that of sodium chloride because the fluoride ion has a smaller radius than the chloride ion
- the lattice energies of the hydroxides or sulfates of group 2 get less exothermic down the group as the radii of the cations get bigger

> **Exam tip**
>
> Do not use the word 'larger' instead of 'more exothermic'. *Numerically* larger is acceptable.

Factors determining hydration enthalpy

Ionic charge

The greater the charge, the stronger is the ion/dipole force of attraction between the ion and the water molecules. Therefore, the hydration enthalpy is more exothermic. For example, the hydration enthalpy of the magnesium ion, Mg^{2+}, is more exothermic (numerically larger) than that of the sodium ion, Na^+.

Ionic radius

The smaller the radius of the ion, the stronger is the force of attraction between the ion and the water molecules. Therefore, the hydration enthalpy is more exothermic (numerically larger). For example:

- the hydration enthalpy of the fluoride ion, F^-, is more exothermic than that of the chloride ion, Cl^-
- the hydration enthalpy of the group 2 cations becomes less exothermic as the group is descended ($Be^{2+} > Mg^{2+} > Ca^{2+} > Sr^{2+} > Ba^{2+}$)

Knowledge check 10

Put the following cations in order of increasingly exothermic hydration enthalpies: Na^+, K^+, Mg^{2+}, Ca^{2+}

Topic 13B Entropy

It is often wrongly assumed that only the value of the enthalpy change determines whether a reaction takes place. There are many examples of processes that are endothermic yet react to completion, and of others that do not react even though a calculated value of ΔH is negative. The direction of change is determined by the thermodynamics of the process.

In any spontaneous process, the *total* entropy change will be positive. This can be written as $\Delta S_{total} > 0$.

Entropy changes in a reaction

The entropy, S, of a substance is determined by how *random* the arrangement and energy of the molecules of that substance are. It is a measure of *disorder*. Thus:

- The entropy of any substance increases with temperature.
- A gaseous substance has a higher entropy than the less randomly arranged liquid form of that substance which, in turn, has a higher entropy than the highly ordered solid, for example, S(water vapour) > S(liquid water) > S(ice).
- A complex substance has a higher entropy than a simple substance in the same physical state. Examples are that the entropy of $CaCO_3$(s) is greater than that of CaO(s), the entropy of PBr_3(l) is greater then that of PCl_3(l) and the entropy of NH_3(g) is greater than that of H_2(g).
- The entropy of the system will increase if the number of moles of product is greater than the number of moles of reactants. Both reactants and products need to be in the same physical state for this to be true.

The entropies of substances will be given in an exam and are quoted under standard conditions of a stated temperature, usually $298\,K$, and $100\,kPa$ pressure.

When any chemical reaction takes place, the entropy of the products will be different from the entropy of the reactants. The difference is called the entropy change of the *system*, ΔS_{system}, and is given by the equation:

$$\Delta S_{system} = \Sigma S_{products} - \Sigma S_{reactants}$$

Worked example

Calculate ΔS_{system} for the reaction of magnesium with oxygen to form magnesium oxide:

$$2Mg(s) + O_2(g) \rightarrow 2MgO(s) \quad \Delta H^{\ominus} = -1203 \, kJ \, mol^{-1}$$

Standard entropies (in $J \, K^{-1} \, mol^{-1}$) at 298 K are Mg(s) 32.7, $O_2(g)$ 205 and MgO(s) 26.9.

Answer

$$\Delta S_{system} = 2 \times S(MgO) - [2 \times S(Mg) + S(O_2)]$$
$$= 2 \times 26.9 - [2 \times 32.7 + 205] = -216.6 \, J \, K^{-1} \, mol^{-1}$$

This is a negative number, but the reaction happens rapidly and completely when the magnesium is ignited. This shows that the entropy change of the *system* is not the only factor that determines whether a reaction is thermodynamically spontaneous or feasible.

What has not yet been considered is the heat given off in this exothermic reaction. This increases the disorder of the surroundings.

The *total* entropy change is made up of two parts: the entropy change of the chemicals, ΔS_{system}, and the entropy change of the surroundings, ΔS_{surr}. This is shown by the equation:

$$\Delta S_{total} = \Delta S_{system} + \Delta S_{surr}$$

ΔS_{system} can be evaluated from data tables as in the worked example above. ΔS_{surr} is calculated from the relationship:

$$\Delta S_{surr} = \frac{-\Delta H}{T}$$

where the temperature T must be in kelvin.

So:

$$\Delta S_{total} = \Delta S_{system} - \frac{\Delta H}{T}$$

Do not forget the minus sign in this expression. Exothermic reactions have a negative ΔH and so a positive ΔS_{surr}.

> **Exam tip**
>
> A physical or chemical change will only be thermodynamically spontaneous (feasible) if ΔS_{total} is positive.

> **Exam tip**
>
> Be aware that entropy values are normally measured in $J \, K^{-1} \, mol^{-1}$ and enthalpy changes are measured in $kJ \, mol^{-1}$. Therefore, the value of ΔS_{surr} must be multiplied by 1000 before being added to ΔS_{system}.

Worked example

Calculate the value of ΔS_{surr} for the reaction between magnesium and oxygen. Use this value and the value of ΔS_{system} from the worked example above to calculate ΔS_{total} and comment on whether the reaction is thermodynamically spontaneous (feasible).

Answer

$$\Delta H^{\ominus} = -1203 \, kJ \, mol^{-1}$$
$$\Delta S_{surr} = \frac{-\Delta H^{\ominus}}{T} = -\left(\frac{-1203 \times 1000}{298}\right) = +4036.9 \, J \, K^{-1} \, mol^{-1}$$
$$\Delta S_{total} = \Delta S_{system} + \Delta S_{surr} = -216.6 + (+4036.9) = +3820 \, J \, K^{-1} \, mol^{-1}$$

The large positive value for ΔS_{total} shows that this reaction is thermodynamically spontaneous and will go to completion. This is in spite of the negative ΔS_{system}.

Table 9 shows whether or not a reaction will be thermodynamically spontaneous (feasible).

ΔH	ΔS_{surr}	ΔS_{system}	Spontaneous?
Negative (exothermic)	Positive	Positive	Always spontaneous
Negative (exothermic)	Positive	Negative	Spontaneous only if ΔS_{surr} outweighs ΔS_{system}
Positive (endothermic)	Negative	Negative	Never spontaneous
Positive (endothermic)	Negative	Positive	Spontaneous only if ΔS_{system} outweighs ΔS_{surr}

Table 9 Spontaneity of reaction

Endothermic reactions that are thermodynamically feasible

- Dissolving ammonium nitrate:

$NH_4Cl(s) + aq \rightarrow NH_4^+(aq) + Cl^-(aq)$ ΔH is positive

The decrease in entropy of the surroundings is more than compensated for by the increase in entropy of the system because an ordered solid becomes disordered ions in solution.

- Reacting solid hydrated barium hydroxide with solid ammonium chloride:

$Ba(OH)_2.8H_2O(s) + 2NH_4Cl(s) \rightarrow BaCl_2(s) + 2NH_3(g) + 10H_2O(l)$ ΔH is positive

The decrease in entropy of the surroundings is more than compensated for by the increase in entropy of the system as 3 moles of ordered solid become 1 mole of ordered solid, 10 moles of less ordered liquid and 2 moles of highly disordered gas.

Free energy change, ΔG

Free energy, G, is a measure of chemical potential. It is defined as:

$G = H - TS$

where H is the enthalpy, T the temperature in kelvin and S the entropy of the system.

For a reaction A $\rightarrow$ B:

$\Delta G = G_B - G_A = (H_B - H_A) - T(S_B - S_A) = \Delta H - T\Delta S$

For a spontaneous change (one that is thermodynamically feasible) ΔG must be negative. This is helped by a negative ΔH and a positive ΔS (Table 10).

ΔH	ΔS	ΔG and hence feasibility
Negative (exothermic)	Positive	ΔG always negative, so always feasible
Negative	Negative	ΔG only negative if $T\Delta S$ is *less* negative than ΔH
Positive (endothermic)	Positive	ΔG only negative if $T\Delta S$ is *more* positive than ΔH
Positive	Negative	ΔG always positive, so never feasible

Table 10 Spontaneous change and signs of ΔH, ΔS and ΔG

Exam tip

The word 'spontaneous' can be misleading because the reaction may not take place for kinetic reasons (activation energy too high). The word 'feasible' can be used instead.

Exam tip

A reaction where ΔS_{total} is positive is thermodynamically feasible and will take place unless the activation energy is too high.

Exam tip

The relationship $\Delta G = \Delta H - T\Delta S$ must be known.

Change of spontaneity

This will happen at a temperature when $\Delta G = 0$.

$$\Delta G = \Delta H - T\Delta S = 0$$

So $T = \dfrac{\Delta H}{\Delta S}$

An endothermic reaction will *become* spontaneous (feasible) at the temperature when

$T = \dfrac{\Delta H}{\Delta S}$

An exothermic reaction will *cease* to be spontaneous when $T = \dfrac{\Delta H}{\Delta S}$

Worked example

Calculate the temperature at which the following reaction becomes feasible:

$$CH_4(g) + H_2O(g) \rightleftharpoons CO(g) + 3H_2(g) \qquad \Delta H = +206\,\text{kJ mol}^{-1}$$

Entropy values/$J\,K^{-1}\,mol^{-1}$: $CH_4(g) = 186$; $H_2O(g) = 189$; $CO(g) = 198$; $H_2(g) = 131$

Answer

$$\Delta S = \Sigma(\Delta S_{products}) - \Sigma(\Delta S_{reactants}) = (198 + 3 \times 131) - (186 + 189) = +216\,\text{J K}^{-1}\,\text{mol}^{-1}$$

$$T = \frac{\Delta H}{\Delta S} = \frac{206\,000\,\text{J mol}^{-1}}{216\,\text{J K}^{-1}\,\text{mol}^{-1}} = 954\,\text{K (or } 681°\text{C)}$$

Relationship between K, ΔG and ΔS_{total}

In terms of ΔG

$$\Delta G = -RT \ln K$$

But:

$$\Delta G = \Delta H - T\Delta S = -RT \ln K$$

$$\ln K = \frac{\Delta S}{R} - \frac{\Delta H}{RT}$$

Worked example

Use the data from the worked example above, to find the value of the equilibrium constant K at $1000\,\text{K}$.

$$CH_4(g) + H_2O(g) \rightleftharpoons CO(g) + 3H_2(g) \qquad \Delta H = +206\,\text{kJ mol}^{-1}$$

Answer

$$\ln K = \frac{\Delta S}{R} - \frac{\Delta H}{RT} = \frac{216/8.31 - (+206\,000)}{(8.31 \times 1000)} = 1.21$$

$$K = e^{1.21} = 3.56$$

Exam tip

K derived from free energy does not have any units.

From ΔS_{total}

$$\Delta S_{total} = R \ln K = \Delta S_{system} - \Delta S_{surr} = \Delta S_{system} - \frac{\Delta H}{T}$$

$$\ln K = \frac{\Delta S_{system}}{R} - \frac{\Delta H}{RT}$$

which is the same as that derived from ΔG.

Effect of temperature on K

$$\ln K = \frac{\Delta S_{system}}{R} - \frac{\Delta H}{RT}$$

If the temperature is increased, the *magnitude* of $\dfrac{-\Delta H}{RT}$ decreases.

For **endothermic reactions** ΔH is positive, so $\dfrac{-\Delta H}{RT}$ is a negative number.

Thus an increase in temperature will make it less negative and so $\ln K$ and hence K will increase.

For **exothermic reactions** ΔH is negative, so $\dfrac{-\Delta H}{RT}$ is a positive number.

Thus an increase in temperature will make it less positive and so $\ln K$ and hence K will decrease.

This is the theoretical backing to Le Châtelier's principle.

Extent of reaction

How far a reaction goes is determined by the magnitude of the equilibrium constant. The value of the entropy change and the equilibrium constant give an estimate of how complete a reaction is at a particular temperature.

The relationship between ΔS_{total}, K and extent of reaction is shown in Table 11.

ΔS_{total}/J K^{-1} mol^{-1}	K	Extent of reaction
> +150	> 10^8	Reaction almost complete
Between +60 and +150	Between 10^3 and 10^8	Reaction favours products
Between +60 and −60	Between 10^3 and 10^{-3}	Both reactants and products in significant quantities
Between −60 and −150	Between 10^{-3} and 10^{-8}	Reaction favours reactants
More negative than −150	< 10^{-8}	Reaction not noticeable

Table 11 Relationship between ΔS_{total}, K and extent of reaction

Kinetic stability

The value of ΔS_{total} or of ΔG predicts whether the reaction is thermodynamically spontaneous, that is, whether it is feasible. Not all reactions that are thermodynamically feasible at a given temperature take place. If the activation energy is too high, too few molecules possess that energy on collision and so the reaction does not take place at that temperature. This is where the reactants are said to be kinetically stable or inert. Magnesium and oxygen at 298 K are thermodynamically unstable relative to magnesium oxide, but the reaction is slow unless they are heated, when the magnesium catches fire. The reaction is said to be thermodynamically feasible but kinetically inert.

The activation energy for the endothermic reaction between ethanoic acid and ammonium carbonate is low and ΔG is negative, so the reaction takes place rapidly and completely at room temperature. The reactants are said to be thermodynamically and kinetically unstable.

Summary

After studying this topic, you should be able to:

- define lattice energy, enthalpy of atomisation and electron affinity
- draw Born–Haber cycles
- estimate the extent of polarisation of an anion from experimental and theoretical lattice energies
- define enthalpy of solution and hydration and use them in Hess's law cycles
- understand the effect of ionic radius and charge on lattice energies and enthalpies of hydration
- define entropy as a measure of disorder

- predict the sign of ΔS_{system} from an equation
- calculate ΔS_{system} from entropy data, ΔS_{surr} from enthalpy data and hence ΔS_{total}
- predict the feasibility of a reaction from the sign of ΔS_{total}
- use the expression for ΔG to predict whether a reaction is feasible and to calculate the value of the equilibrium constant
- explain the effect of a change in temperature on ΔS_{total} and hence on ΔG
- explain why a reaction that is thermodynamically spontaneous may not take place

■ Topic 14 Redox II

Required year 1 knowledge

Definitions

- **Oxidation** occurs when an atom, molecule or ion *loses* one or more electrons. The oxidation number of the element involved increases.
- **Reduction** is when an atom, molecule or ion *gains* one or more electrons. The oxidation number of the element involved decreases.
- An **oxidising agent** is a substance that oxidises another substance and is itself reduced. The oxidation number of an element in the oxidising agent decreases.
- A **reducing agent** is a substance that reduces another substance and is itself oxidised. Its oxidation number increases.
- The **oxidation number** of an atom in a compound is the charge that it would have if the compound were totally ionic.

Exam tip

Remember **OILRIG** — **O**xidation **I**s **L**oss, **R**eduction **I**s **G**ain.

Oxidation number

The rules for working out oxidation numbers should be applied in the following order:

- Rule 1 — the oxidation number of an element in its standard state is zero.
- Rule 2 — a simple monatomic ion has an oxidation number equal to its charge.
- Rule 3 — the sum of the oxidation numbers in a neutral formula is zero.
- Rule 4 — the sum of the oxidation numbers in an ion adds up to the charge on that ion.

Knowledge check 11

What are the oxidation numbers of
a chromium in $Cr_2O_7^{2-}$ and **b** iodine in IO_3^-?

Worked example

What is the oxidation number of iron in Fe_3O_4?

Answer

The oxidation state of each oxygen is -2. Therefore, the four oxygen atoms are $4 \times -2 = -8$. The three iron atoms are $+8$ in total. Thus, the average oxidation state of each iron atom is $\frac{8}{3} = 2\frac{2}{3}$.

The answer is not a whole number because two of the iron atoms are in the $+3$ state and one is in the $+2$ state.

Ionic half-equations

Ionic half-equations always have electrons on either the left-hand side or the right-hand side.

When zinc is added to dilute hydrochloric acid, the hydrogen ions are reduced to hydrogen, because their oxidation number decreases from $+1$ to zero. The half-equation is:

$$2H^+(aq) + 2e^- \rightleftharpoons H_2(g)$$

As this is a reduction reaction, the electrons are on the left-hand side of the half-equation.

The zinc atoms are oxidised to zinc ions, because their oxidation number increases from zero to $+2$:

$$Zn(s) \rightleftharpoons Zn^{2+}(aq) + 2e^-$$

As this is oxidation, the electrons are on the right-hand side of the equation.

Change in oxidation number and reaction stoichiometry

The total increase in oxidation number of the element in the reducing agent must equal the total decrease in oxidation number of the element in the oxidising agent. Note that a reducing agent gets *oxidised* and an *oxidising* agent gets reduced.

Worked example

Bromate ions (BrO_3^-) react with bromide ions (Br^-), in acidic solution to form bromine and water. Evaluate the oxidation numbers of the bromine species and hence work out the reaction stoichiometry. Write the overall equation.

Answer

The oxidation numbers in BrO_3^- add up to -1. Each oxygen is -2, so the three oxygen atoms are $3 \times -2 = -6$. The bromine in BrO_3^- is $+5$, because $+5 + (-6) = -1$.

The oxidation number of bromine in Br^- is -1.

The oxidation number of bromine in Br_2 is zero.

→

Exam tip

If the question mentions 'in acidic solution', there will be H^+ ions on the left in the overall equation.

As the oxidation number of bromine in BrO_3^- decreases by five and the bromine in Br^- increases by one, there must be $5Br^-$ to each BrO_3^-.

Therefore the equation is:

$$5Br^- + BrO_3^- + 6H^+ \rightarrow 3Br_2 + 3H_2O$$

Knowledge check 12

Calculate the oxidation numbers of manganese and carbon in the redox equation below between manganate ions and ethanedioate ions and use them to complete the equation.

$$...MnO_4^- + ...C_2O_4^{2-} + ...H^+ \rightarrow ...Mn^{2+} + ...CO_2 + ...H_2O$$

Year 2 redox

Standard electrode potential, $E^\ominus$

This is defined as the electrical potential of an *electrode* relative to a standard hydrogen electrode, where all ions in solution are at $1\,mol\,dm^{-3}$ concentration and any gases are at $100\,kPa$ pressure and a stated (usually 25°C) temperature.

The values are always stated as *reduction* potentials, with electrons on the left:

$$Zn^{2+}(aq) + 2e^- \rightleftharpoons Zn(s) \qquad E^\ominus = -0.76\,V$$

so the standard electrode potential is also called the standard *reduction* potential. The data can be given as a reduction equation, as above, or as $E^\ominus\,(Zn^{2+}/Zn) = -0.76\,V$.

Exam tip

You will need to use the *Data Booklet* to find standard electrode potentials. The table gives the species in *increasing* power as oxidising agents. Thus iodine is a weaker oxidising agent than bromine.

Standard hydrogen electrode

This is the reference electrode in all electrochemical measurements. Its value at 25°C is defined as zero volts. It consists of hydrogen gas at $100\,kPa$ pressure being passed over a platinum electrode dipping into a solution of $1.0\,mol\,dm^{-3}$ H^+ ions (pH = 0) at 25°C:

$$H^+(aq) + e^- \rightleftharpoons \tfrac{1}{2}H_2(g) \qquad E^\ominus = 0.00\,V$$

A reference electrode is needed, as a potential cannot be measured on its own — only a potential difference.

Measurement of standard electrode potentials

The voltage of the electrode or half-cell is measured relative to a standard, such as the standard hydrogen electrode. For zinc, the half-equation is $Zn^{2+}(aq) + 2e^- \rightleftharpoons Zn(s)$. Thus, it is the potential of a zinc rod dipping into a $1.0\,mol\,dm^{-3}$ solution of Zn^{2+} ions at 298 K, relative to a standard hydrogen electrode.

For chlorine, the half-equation is $Cl_2(g) + 2e^- \rightleftharpoons 2Cl^-(aq)$. Thus, it is the potential of chlorine gas, at $100\,kPa$ pressure, bubbled over a platinum electrode dipping into a $1.0\,mol\,dm^{-3}$ solution of Cl^- ions at 298 K, relative to a standard hydrogen electrode.

For the Fe^{3+}/Fe^{2+} system, the half-equation is $Fe^{3+}(aq) + e^- \rightleftharpoons Fe^{2+}(aq)$. Thus, it is the potential when a platinum electrode is placed into a solution of concentration

Exam tip

When measuring a standard electrode potential, all soluble substances, whether they are on the left or the right of the half-equation, must be at a concentration of $1.0\,mol\,dm^{-3}$.

$1.0\,\text{mol dm}^{-3}$ in *both* Fe^{3+} and Fe^{2+} ions at 298 K, relative to a standard hydrogen electrode (Figure 7).

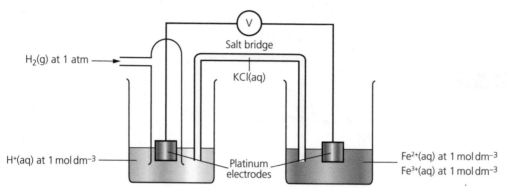

Figure 7

The right-hand electrode system is one of:

- a metal dipping into a solution of its ions at $1\,\text{mol dm}^{-3}$
- a platinum electrode with the gaseous non-metal at a pressure of $100\,\text{kPa}$ bubbling over it dipping into a solution of its ions at a concentration of $1\,\text{mol dm}^{-3}$
- a platinum electrode dipping into a solution containing ions of the same element in different oxidation states (e.g. Fe^{2+} and Fe^{3+}) both at a concentration of $1\,\text{mol dm}^{-3}$

Knowledge check 13

Describe the electrode that is linked to a standard hydrogen electrode when measuring the standard electrode potential of the $[MnO_4^- + H^+, Mn^{2+} + H_2O]$ system, giving the conditions at that electrode.

Core practical 10

Investigating some electrochemical cells

A standard hydrogen electrode is difficult to use in a school laboratory, so a secondary standard is often used instead. This is a calomel electrode, which has $E = +0.27\,\text{V}$. It consists of a mercury electrode dipping into a saturated solution of mercury(I) chloride and potassium chloride.

To find the standard electrode potential of zinc:

1 Put some saturated potassium chloride in one beaker and a $1.00\,\text{mol dm}^{-3}$ solution of zinc sulfate in another, making sure that the levels are the same.

2 Connect the two beakers with a strip of filter paper soaked in saturated potassium nitrate.

3 Place a clean zinc rod into the beaker containing zinc sulfate and the calomel electrode into the other beaker.

4 Connect the two electrodes via a high resistance voltmeter and read the voltage generated by the cell.

5 The reading was $+1.03\,\text{V}$:

$E_{cell} = E^\ominus \text{(calomel)} - E^\ominus \text{(Zn}^{2+}\text{/Zn)}$

$E^\ominus \text{(Zn}^{2+}\text{/Zn)} = E^\ominus \text{(calomel)} - E_{cell} = +0.27 - 1.03 = -0.76\,\text{V}$

The experiment can be repeated with different electrode systems, each time using the calomel electrode as a secondary standard.

Cell diagrams

The cell diagram is a standard way of representing cells and redox processes. The diagram is written in the following order, from left to right:

- the formula of the metallic anode followed by a vertical line
- the ions present in the anode area followed by two vertical lines
- the ions present in the cathode area followed by a vertical line
- the formula of the metallic cathode

For the cell where zinc is oxidised (so is the anode) and copper ions reduced to copper (which is the cathode), the cell diagram is:

$$Zn(s) \mid Zn^{2+}(aq) \parallel Cu^{2+}(aq) \mid Cu(s)$$

For the cell where manganate(VII) ions oxidise iron(II) ions (so MnO_4^- is reduced and Fe^{2+} is oxidised), the cell diagram is:

$$Pt(s) \mid MnO_4^-(aq), 8H^+(aq), Mn^{2+}(aq) \parallel Fe^{3+}(aq), Fe^{2+}(aq) \mid Pt(s)$$

Calculation of $E^\ominus_{cell}$ for a reaction and hence feasibility

A redox reaction will be feasible if E_{cell} is positive.

If a question asks, 'Will substance A oxidise substance B?', you need to look for the half-equations containing A and B. You will find A on the left and B on the right of their half-equations.

There are three methods for calculating $E^\ominus_{cell}$ (which is also called $E^\ominus_{reaction}$).

Method 1

This is based on the formula:

$$E^\ominus_{cell} = E^\ominus \text{ (oxidising agent)} - E^\ominus \text{ (reducing agent)}$$

where $E^\ominus$ refers to the standard reduction potential, i.e. the half-equation with electrons on the *left*.

- Step 1: Identify the oxidising agent. Remember that the oxidising agent is reduced and so is on the left of a standard electrode (reduction) potential equation.
- Step 2: Identify the reducing agent. The reducing agent is oxidised and so is on the right of a standard electrode (reduction) potential.
- Step 3: Calculate $E^\ominus_{cell} = E^\ominus$ of oxidising agent $- E^\ominus$ of reducing agent, where the $E^\ominus$ values are the standard *reduction* potentials of the oxidising agent (on the left in its half-equation) and of the reducing agent (on the right in its half-equation).

  Thus, a reaction will be feasible if the $E^\ominus$ (oxidising agent) is more positive (or less negative) than the $E^\ominus$ of the half-equation with the reducing agent on the right.

Exam tip

Remember that a redox reaction is feasible if $E^\ominus_{cell}$ is positive.

Will manganate(VII) ions in acid solution oxidise chloride ions to chlorine?

$$Cl_2 + 2e- \rightleftharpoons 2Cl \qquad\qquad E^\ominus = +1.36\,V$$

$$MnO_4^- + 8H^+ \rightleftharpoons Mn^{2+} + 4H_2 \qquad E^\ominus = +1.52\,V$$

Answer

$$E^\ominus{}_{cell} = E^\ominus \text{ (oxidising agent)} -E^\ominus \text{ (reducing agent)}$$

$$= E^\ominus \text{ } (MnO_4^-/Mn^{2+}) - E^\ominus \text{ } (Cl_2/Cl^-)$$

$$= +1.52 - (+1.36) = +0.16\,V$$

This is a positive number, so manganate(VII) ions, in acid solution, will oxidise chloride ions. (The reaction is thermodynamically spontaneous.)

Method 2: the anticlockwise rule

If the question asks 'Will A oxidise B?', write the two electrode systems with the more negative above the less negative (or the less positive above the more positive). Work in an anticlockwise direction starting on the left of the bottom equation.

Use the *Data Booklet* to see if hydrogen peroxide will oxidise chromium 3+ ions to dichromate(VI) ions.

Answer

Place them with the less positive above the more positive.

$$[Cr_2O_7{}^{2-}(aq) + 14H^+(aq)], [2Cr^{3+}(aq) + 7H_2O(l)] \quad E^\ominus = +1.33\,V$$

$$[H_2O_2(aq) + 2H^+(aq)], 2H_2O(l) \qquad\qquad\qquad E^\ominus = +1.77\,V$$

Move anticlockwise from H_2O_2.

H_2O_2 (in acid solution) will oxidise Cr^{3+} ions to $Cr_2O_7{}^{2-}$ and the value of $E^\ominus{}_{cell} = +1.77 - (+1.33) = +0.44\,V$.

As this is positive, the reaction is thermodynamically feasible.

Use the data below to explain whether IO_3^- ions will oxidise NO.

$$½N_2O_4 + 2H^+ + 2e^- \rightleftharpoons NO + H_2O \qquad E^\ominus = +1.03\,V$$

$$IO_3^- + 6H^+ + 5e^- \rightleftharpoons ½I_2 + 3H_2O \qquad E^\ominus = +1.19\,V$$

The reactants will be on the left of the bottom electrode system and the right of the top one. The products will be on the right of the bottom electrode system and the left of the top one.

Method 3

The value of $E^{\ominus}_{cell}$ can also be obtained by reversing the half-equation containing substance B and at the same time changing the sign of its $E^{\ominus}$ value. Then add it to the half-equation for substance A. This is the best method if the overall equation is also required.

Worked example 3

Will iron(III) ions oxidise iodide ions to iodine?

$$Fe^{3+} + e^- \rightleftharpoons Fe^2 \quad E^{\ominus} = +0.77\,V$$

$$\tfrac{1}{2}I_2 + e^- \rightleftharpoons I^- \quad E^{\ominus} = +0.54\,V$$

Answer

The reactants are Fe^{3+} (on the left of its half-equation) and I^- (on the right of its half-equation).

The I_2/I^- half-equation is reversed, its sign changed and then it is added to the Fe^{3+}/Fe^{2+} half-equation.

$$I^- \rightarrow \tfrac{1}{2}I_2 + e^- \quad E^{\ominus} = -(+0.54)\,V = -0.54\,V$$

$$Fe^{3+} + e^- \rightarrow Fe^{2+} \quad E^{\ominus} = +0.77\,V$$

Adding gives the overall equation and the value of $E^{\ominus}_{cell}$:

$$Fe^{3+} + I^- \rightarrow Fe^{2+} + \tfrac{1}{2}I_2 \quad E^{\ominus}_{cell} = +0.77\,V + (-0.54\,V) = +0.23\,V$$

This is a positive number, so the reaction is feasible and iron(III) ions oxidise iodide ions.

Exam tip

If you reverse a half-equation, you must alter its sign. However, if you multiply it by a number, you do *not* alter its value.

Extent of reaction

A reaction is said to be feasible (thermodynamically spontaneous) if the value of the cell potential ($E^{\ominus}_{cell}$) is *positive*. The extent of the reaction depends on the number of electrons involved and the numerical value of $E^{\ominus}_{cell}$.

$E^{\ominus}_{cell}$ is directly proportional to ΔS_{total} and ΔS_{total} is connected to the equilibrium constant by the expression:

$$\Delta S_{total} = R \ln K$$

So E_{cell} is directly proportional to $\ln K$. This means that a positive E_{cell} results in a positive ΔS_{total} and hence a value of K greater than 1.

A negative value of E_{cell} results in a negative ΔS_{total} and a value of K that is less than 1. The more positive E_{cell}, the larger the value of ΔS_{total} and the larger the value of the equilibrium constant, K. Thus, large positive values of E_{cell} result in a virtually complete reaction and small values result in an incomplete reaction.

Writing overall redox equations

Questions often provide the necessary reduction half-equations and ask you to write an overall equation. This is done in four steps:

- Step 1: Look at the overall reaction and identify the *reactants* in the two half-equations or electrode systems. One reactant (the oxidising agent) is on the left-hand side and the other reactant (the reducing agent) is on the right-hand side.
- Step 2: Reverse the equation that has the reactant on the *right*.
- Step 3: Multiply one or both equations, so that both now have the same number of electrons. One equation will still have the electrons on the left and the other will now have them on the right.
- Step 4: Add the two equations. *Cancel* the electrons, as there will be the same number on each side of the equation.

Worked example

Write the overall equation for the oxidation of Sn^{2+} ions by manganate(VII) ions (MnO_4^-) in acid solution and predict whether it will be thermodynamically spontaneous. The two reduction half-equations are:

$$Sn^{4+} + 2e^- \rightleftharpoons Sn^{2+} \qquad\qquad E^\ominus = +0.15\,V$$

$$MnO_4^- + 8H^+ + 5e^- \rightleftharpoons Mn^{2+} + 4H_2O \qquad E^\ominus = +1.52\,V$$

Answer

Step 1: the reactants are MnO_4^- and Sn^{2+}.

Step 2: reverse the equation with the reactant on the right (the first half-equation):

$$Sn^{2+} \rightleftharpoons Sn^{4+} + 2e^- \qquad\qquad E^\ominus = -(+0.15) = -0.15\,V$$

Step 3: multiply the reversed first equation by five and the second equation by two so that both half-equations have the same number of electrons:

$$5Sn^{2+} \rightleftharpoons 5Sn^{4+} + 10e^- \qquad E^\ominus = -0.15\,V$$

$$2MnO_4^- + 16H^+ + 10e^- \rightleftharpoons 2Mn^{2+} + 8H_2O \qquad E^\ominus = +1.52\,V$$

Step 4: add to give the overall equation:

$$5Sn^{2+} + 2MnO_4^- + 16H^+ \rightleftharpoons 5Sn^{4+} + 2Mn^{2+} + 8H_2O$$

The value of $E^\ominus_{\text{reaction}}$ is found by adding the two $E^\ominus$ values at step 3:

$$E^\ominus_{\text{reaction}} = +1.52 + (-0.15\,V) = +1.37\,V$$

This is a positive number and so the reaction is thermodynamically spontaneous (feasible).

Exam tip

When a half-equation is reversed, the *sign* of its $E^\ominus$ value is changed.

Exam tip

Note that when a half-equation is multiplied by a number, its value is *not* altered.

Knowledge check 15

Use the *Data Booklet* and $HCl + H^+ + e^- \rightleftharpoons H_2O + \frac{1}{2}Cl_2$ $E^\ominus = +1.59\,V$ to predict whether chloric(I) acid, HClO, will oxidise chloride ions, Cl^- in acid solution and, if so, write the overall equation.

When predictions and experiment do not agree

There could be several reasons for this.

Kinetic reasons

A reaction that is thermodynamically spontaneous (positive $E^\ominus_{cell}$) may not take place if the activation energy is too high.

An example of this is the oxidation of ethanedioate ions by manganate(VII) ions. The value of $E^\ominus_{cell}$ is +2.01 V, and so the reaction is classified as being thermodynamically spontaneous. But this reaction does not happen at room temperature, as the activation energy is too high. On warming, the reaction proceeds quickly enough for the two substances to be titrated.

Concentration effects: non-standard conditions

Aqueous copper(II) ions should not be reduced to aqueous copper(I) ions by iodide ions:

$$Cu^{2+}(aq) + e^- \rightleftharpoons Cu^+(aq) \qquad E^\ominus = +0.15\,V$$

$$I_2(aq) + 2e^- \rightleftharpoons 2I^-(aq) \qquad E^\ominus = +0.54\,V$$

Copper(II) ions are the oxidising agent and iodide ions the reducing agent, so for the reaction:

$$Cu^{2+}(aq) + I^-(aq) \rightleftharpoons Cu^+(aq) + \tfrac{1}{2}I_2(aq)$$
$$E^\ominus_{cell} = +0.15 - (+0.54) = -0.39\,V \quad \text{(equation 1)}$$

$E^\ominus_{cell}$ is negative, so this reaction is *not* thermodynamically spontaneous under standard conditions. However, when aqueous copper(II) and iodide ions are mixed, the iodide ions do reduce copper(II) ions to copper(I). The reason can be seen in the equation for the reaction that then happens:

$$Cu^+(aq) + I^-(aq) \rightarrow CuI(s) \qquad \text{(equation 2)}$$

Thus the overall reaction is:

$$Cu^{2+}(aq) + 2I^-(aq) \rightarrow CuI(s) + \tfrac{1}{2}I_2(aq)$$

Copper(I) iodide is very insoluble and so $[Cu^+(aq)]$ is close to zero. This drives the equilibrium for equation 1 over to the right. The conditions are not standard, so that although $E^\ominus_{cell}$ is negative, E_{cell} for the *non-standard* conditions is positive.

Disproportionation

This is a reaction in which an *element* in a *single* species is simultaneously oxidised and reduced. In the following reaction, the chlorine in the species NaOCl is oxidised from +1 to +5 and simultaneously reduced from +1 to −1:

$$3NaOCl \rightarrow NaClO_3 + 2NaCl$$
$$\quad\;+1 \qquad\quad +5 \qquad\;\; -1$$

The half-equations are:

$$OCl^-(aq) + 2H^+(aq) + 2e^- \rightleftharpoons Cl^-(aq) + H_2O(l) \quad E^\ominus = +1.59\,V \quad \text{(equation 1)}$$

> **Exam tip**
>
> This is also an example of autocatalysis, as Mn^{2+} ions, which are produced when the manganate(VII) ions are reduced, are the catalyst for this redox reaction.

$$ClO_3^- + 4H^+ + 3e^- \rightleftharpoons OCl^- + 2H_2O \qquad E^\ominus = +1.45\,V \qquad \text{(equation 2)}$$

$$E^\ominus{}_{cell} = \text{equation 1} - \text{equation 2} = +1.59 - (-1.45) = +0.14\,V$$

This is positive, so disproportionation will occur.

The following reaction is *not* a disproportionation reaction, even though the chlorine in Cl^- has been oxidised and the chlorine in OCl^- has been reduced, because the chlorine is in two *different* species:

$$Cl^- + OCl^- + 2H^+ \rightarrow Cl_2 + H_2O$$
$$-1 \quad\; +1 \qquad\qquad\; 0$$

Storage and fuel cells

Rechargeable battery

The lead–acid battery is used in cars. Each cell consists of two lead plates. The cathode is coated with solid lead(IV) oxide. The electrolyte is a fairly concentrated solution of sulfuric acid.

The anode reaction is:

$$Pb(s) + SO_4^{2-}(aq) \rightarrow PbSO_4(s) + 2e^- \qquad E = +0.34\,V$$

The cathode reaction is:

$$PbO_2(s) + 4H^+(aq) + SO_4^{2-}(aq) + 2e^- \rightarrow PbSO_4(s) + 2H_2O(l) \qquad E = +1.66\,V$$

The overall discharging reaction is:

$$Pb(s) + PbO_2(s) + 4H^+(aq) + 2SO_4^{2-}(aq) \rightarrow 2PbSO_4(s) + 2H_2O(l) \quad E_{cell} = 2.0\,V$$

Usually there are six cells, generating 12 V.

When all the lead(IV) oxide has been reduced, the battery is flat.

The discharging reaction is reversible. If an external potential greater than 12 V is applied, the reaction is driven backwards and the plates restored to their original composition.

The overall charging reaction is:

$$2PbSO_4(s) + 2H_2O(l) \rightarrow Pb(s) + PbO_2(s) + 4H^+(aq) + 2SO_4^{2-}(aq)$$

Hydrogen fuel cells

Manned spacecraft are powered by hydrogen–oxygen fuel cells. These cells are also being developed for commercial use.

Hydrogen gas is oxidised at the anode and the electrons released go round the external circuit powering the machinery to the cathode, where oxygen gas is reduced. The ions produced pass through the electrolyte. The only product is water, which must constantly be removed from the cell.

The electrodes are made of porous graphite embedded with the catalyst, which is usually platinum. There are two distinct types of hydrogen fuel cell.

Knowledge check 16

Write the equation for the disproportionation reaction of manganate(VI) ions, MnO_4^{2-} to manganese(IV) oxide, MnO_2 and manganese(VII) ions, MnO_4^-, in acid solution.

Acid fuel cells

At the anode, hydrogen is oxidised and H^+ ions and electrons are formed. The H^+ ions then pass through the solid polymer electrolyte to the cathode. The electrons released pass round the external circuit to the cathode. Oxygen gas (from the air) is reduced at the cathode and water is formed. The half-equations are:

Anode	$2H_2 \rightleftharpoons 4H^+ + 4e^-$
Cathode	$O_2 + 4H^+ + 4e^- \rightleftharpoons 2H_2O$
Overall	$2H_2 + O_2 \rightarrow 2H_2O$

Alkaline fuel cells

The electrodes are porous platinum and the electrolyte is an aqueous solution of potassium hydroxide. The principle is the same as with acid cells. The hydrogen is oxidised at the anode:

$$2H_2(g) + 4OH^-(aq) \rightleftharpoons 4H_2O(l) + 4e^-$$

The electrons travel around the external circuit to the cathode, where oxygen is reduced:

$$O_2(g) + 2H_2O(l) + 4e^- \rightleftharpoons 4OH^-(aq)$$

Overall:

$$2H_2 + O_2 \rightarrow 2H_2O$$

Alcohol fuel cells

Research is being carried out into designing fuel cells that will use fuels rich in hydrogen. The two half-equations for methanol are:

$$CH_3OH(l) + H_2O(l) \rightleftharpoons HCOOH(aq) + 4H^+ + 4e^-$$

$$O_2(g) + 4H^+(aq) + 4e^- \rightleftharpoons 2H_2O(l)$$

Redox titrations

Estimating the concentration of a solution of an oxidising agent

The general method is to add a known volume of a solution of the oxidising agent from a pipette to an *excess* of acidified potassium iodide and then titrate the liberated iodine with standard sodium thiosulfate solution. Starch is added when the iodine has faded to a pale straw colour, and the sodium thiosulfate solution is then added drop by drop until the blue-black starch–iodine colour disappears.

The equation for the titration reaction is:

$$I_2 + 2Na_2S_2O_3 \rightarrow 2NaI + Na_2S_4O_6$$

This is an equation that you must know and you must be able to use its stoichiometry to calculate the amount of iodine that reacted with the sodium thiosulfate in the

> **Exam tip**
>
> If the starch is added too early, an insoluble starch/iodine complex is formed and the titre will be inaccurate.

titre. For example, if the iodine produced required $12.3\,cm^3$ of $0.456\,mol\,dm^{-3}$ sodium thiosulfate solution, then:

amount (moles) of $Na_2S_2O_3 = 0.456\,mol\,dm^{-3} \times 0.0123\,dm^3 = 0.005\,609\,mol$

amount (moles) of $I_2 = \frac{1}{2} \times 0.005\,609 = 0.002\,80\,mol$

Exam tip

Remember that moles of solute = concentration (in $mol\,dm^{-3}$) × volume (in dm^3).

Core practical 11A

Finding the concentration of an oxidising agent

1 Pipette $25.00\,cm^3$ of a solution of iron(III) chloride into a conical flask.

2 Add excess potassium iodide crystals. (This liberates iodine.)

3 Fill a burette with a standard solution of sodium thiosulfate.

4 Add the sodium thiosulfate slowly from the burette until the iodine has faded to a pale straw colour.

5 Add a few drops of starch and continue to add the thiosulfate until one drop turns the blue-black solution colourless.

6 Read the burette and repeat until at least two consistent titres have been obtained.

Data
The mean titre was $24.20\,cm^3$.

Calculation

Redox equation: $2FeCl_3 + 2KI \rightarrow I_2 + 2FeCl_2 + 2KCl$

Titration equation: $I_2 + 2Na_2S_2O_3 \rightarrow 2NaI + Na_2S_4O_6$

Amount of thiosulfate = $0.100\,mol\,dm^{-3} \times 24.20/1000\,dm^3 = 0.002\,420\,mol$

Amount of iodine = $\frac{1}{2} \times 0.002\,420 = 0.001\,210\,mol$ ($\frac{1}{2}$ because 1 mol of I_2 reacts with 2 mol of $Na_2S_2O_3$)

Amount of $FeCl_3$ = $2 \times 0.001\,210 = 0.002\,420\,mol$ (2× because 2 mol of $FeCl_3$ produce 1 mol of I_2)

Concentration of the $FeCl_3$ solution = $\dfrac{0.002\,420\,mol}{0.025\,00\,dm^3} = 0.0968\,mol\,dm^{-3}$

Knowledge check 17

Iodate(V) ions react with iodide ions according to the equation:

$IO_3^- + 5I^- + 6H^+ \rightarrow 3I_2 + 3H_2O$

$0.87\,g$ of a group 1 iodate, MIO_3, was dissolved in water and made up to $250\,cm^3$. Excess acidified potassium iodide was added to $25\,cm^3$ portions and the liberated iodine was titrated with $0.100\,mol\,dm^{-3}$ sodium thiosulfate. The mean titre was $26.45\,cm^3$. Calculate the molar mass of MIO_3 and hence deduce the identity of the metal M.

Core practical 11B

Finding the concentration of a reducing agent

Almost all reducing agents reduce potassium manganate(VII) in acid solution.

1 Transfer 25.00 cm^3 of the solution of the reducing agent from a pipette into a conical flask.

2 Acidify with approximately 25 cm^3 of dilute sulfuric acid.

3 Rinse a burette with water and then with some of a standard solution of potassium manganate(VII).

4 Add the potassium manganate(VII) solution from the burette until a faint, permanent, pink colour is obtained.

5 Repeat until two consistent titres are obtained and find the average of their values.

6 Calculate the result as in the worked example below.

Exam tip

The colour of potassium manganate(VII) is so intense that there is no need to add an indicator.

Worked example

'Iron' tablets contain iron(II) sulfate. The percentage of iron in these tablets can be found as follows:

■ Weigh one iron tablet.

■ Crush the tablet and dissolve it in about 25 cm^3 of dilute sulfuric acid.

■ Use distilled water to make the solution up to 250 cm^3.

■ Titrate 25.00 cm^3 portions with standard 0.0202 mol dm^{-3} potassium manganate(VII) solution until two consistent titres have been obtained.

Data

Mass of tablet = 10.31 g

Mean titre = 26.20 cm^3

Equation: $MnO_4^- + 8H^+ + 5Fe^{2+} \rightarrow Mn^{2+} + 4H_2O + 5Fe^{3+}$

Answer

Amount of MnO_4^- = 0.0202 mol $dm^{-3} \times \dfrac{26.20\,dm^3}{1000}$ = 0.000 529 2 mol

Amount of Fe^{2+} in 25.00 cm^3 of solution = 5 × 0.000 529 2 = 0.002 646 mol

Amount of Fe^{2+} in 250 cm^3 of solution = 10 × 0.002 646 = 0.026 46 mol

Mass of iron in one tablet = moles × molar mass of Fe = 0.026 46 mol × 55.8 g mol^{-1}

$$= 1.476\,g$$

Percentage of iron in one tablet = $\dfrac{1.476}{10.31} \times 100$ = 14.3%

Knowledge check 18

Rhubarb leaves contain ethanedioic acid, $H_2C_2O_4$, which is poisonous. 3.14 g of rhubarb leaves were crushed in dilute acid and the ethanedioic acid liberated was titrated with $0.0100 \, mol \, dm^{-3}$ potassium manganate(VII) solution. The titre was $16.80 \, cm^3$. Calculate the percentage of ethanedioic acid in the rhubarb leaves. The equation for the reaction is:

$$2MnO_4^- + 5H_2C_2O_4 + 6H^+ \rightarrow 2Mn^{2+} + 10CO_2 + 8H_2O$$

Summary

Make sure that you know:
- the definition of oxidation and reduction in terms of electron transfer and oxidation number
- what is meant by standard electrode potential and how to measure them
- how to describe a hydrogen electrode
- how to calculate $E^\ominus_{cell}$ and predict feasibility
- that $E^\ominus_{cell}$ is directly proportional to ΔS_{total} and to $\ln K$
- how to explain why $E^\ominus_{cell}$ might predict a reaction but it is not observed
- how to describe storage and fuel cells
- how to carry out calculations given redox titration data

■ Topic 15 Transition metals

Topic 15A Principles of transition metal chemistry

Definitions

- **d-block elements** are those in which the last electron has gone into a *d*-orbital.
- A **transition element** has partially filled *d*-orbitals in one or more of its ions. Transition elements are all *d*-block elements. All are metals; most are physically strong and have high melting points.
- A **ligand** is a molecule or negative ion that uses a lone pair of electrons to form a dative bond with a *d*-block cation.
- The **coordination number** is the number of atoms, ions or groups datively bonded to a *d*-block cation.
- A **dative (coordinate) bond** is formed when two atoms share a pair of electrons, both of which are supplied by one atom.

Exam tip

When defining a *d*-block element, do not say 'outermost electrons are in *d*-orbitals', or 'the highest energy electron is in a *d*-orbital', because neither is true.

Electron configuration

The electron configurations of the d-block elements are shown in Table 12.

Element	Electron configuration	3d	4s
Sc	[Ar] $3d^1\,4s^2$	↑	↑↓
Ti	[Ar] $3d^2\,4s^2$	↑ ↑	↑↓
V	[Ar] $3d^3\,4s^2$	↑ ↑ ↑	↑↓
Cr	[Ar] $3d^5\,4s^1$	↑ ↑ ↑ ↑ ↑	↑
Mn	[Ar] $3d^5\,4s^2$	↑ ↑ ↑ ↑ ↑	↑↓
Fe	[Ar] $3d^6\,4s^2$	↑↓ ↑ ↑ ↑ ↑	↑↓
Co	[Ar] $3d^7\,4s^2$	↑↓ ↑↓ ↑ ↑ ↑	↑↓
Ni	[Ar] $3d^8\,4s^2$	↑↓ ↑↓ ↑↓ ↑ ↑	↑↓
Cu	[Ar] $3d^{10}\,4s^1$	↑↓ ↑↓ ↑↓ ↑↓ ↑↓	↑
Zn	[Ar] $3d^{10}\,4s^2$	↑↓ ↑↓ ↑↓ ↑↓ ↑↓	↑↓

Table 12 Electron configurations of d-block metals

All the elements have the configuration [Ar] $3d^x\,4s^2$, except chromium ([Ar] $3d^5\,4s^1$) and copper ([Ar] $3d^{10}\,4s^1$). The difference is caused by the extra stability of a half-filled or fully filled d sub-shell, which makes it energetically preferable for an electron to move out of the $4s$-orbital into a $3d$-orbital. An atom of iron has six $3d$-electrons because it is the sixth d-block element. The electron configurations of atomic iron, the Fe^{2+} ion and the Fe^{3+} ion are shown in Table 13.

Element	Electron configuration	3d	4s
Fe	[Ar] $3d^6\,4s^2$	↑↓ ↑ ↑ ↑ ↑	↑↓
Fe^{2+}	[Ar] $3d^6$	↑↓ ↑ ↑ ↑ ↑	
Fe^{3+}	[Ar] $3d^5$	↑ ↑ ↑ ↑ ↑	

Table 13 Electron configuration of states of iron

When a transition metal loses electrons and forms a cation, the first electrons lost come from the $4s$-orbital. Any further electrons come from paired electrons in d-orbitals. Thus the Fe^{3+} ion has no $4s$-electrons and five unpaired $3d$-electrons.

Evidence for electron configurations

There is a significant jump in the successive ionisation energies between removing the last $4s$-electron and the first $3d$-electron. For chromium and copper, this jump is between the first and second ionisation energies. For the other d-block elements, it lies between the second and third ionisation energies.

Exam tip

[Ar] is short for the electron configuration of argon, which is $1s^2\,2s^2\,2p^6\,3s^2\,3p^6$.

Knowledge check 19

Write the electron configuration for
a the Cr^{3+} ion and
b the Mn^{2+} ion.

Exam tip

All d-block cations have a $4s^0$ electron configuration.

There is also a big jump after the last $3d$-electron has been removed, as the next electron comes from a $3p$-orbital that is much more strongly held. For vanadium, this is after the fifth ionisation energy. For chromium, it is after the sixth; for manganese after the seventh; and for iron after the eighth ionisation energy. This is the evidence for chromium and copper having only one $4s$-electron, and for the total number of $4s$- and $3d$-electrons.

Variable oxidation state

Transition elements have several different oxidation states. The common oxidation states of the d-block metals are shown in Table 14.

Sc	Ti	V	Cr	Mn	Fe	Co	Ni	Cu	Zn
								+1	
	+2	+2	+2	+2	+2	+2	+2	+2	+2
+3	+3	+3	+3		+3	+3			
	+4	+4		+4					
		+5							
			+6	+6					
				+7					

Table 14 Oxidation states of d-block metals

- Chromium can be:
  - Cr^{2+}, e.g. $CrCl_2$
  - Cr^{3+}, e.g. $Cr_2(SO_4)_3$
  - in the +6 state, e.g. K_2CrO_4 and $K_2Cr_2O_7$
- Iron can be:
  - Fe^{2+}, e.g. $FeSO_4$
  - Fe^{3+}, e.g. $FeCl_3$
- Copper can be:
  - Cu^+, e.g. $CuCl$ and Cu_2O
  - Cu^{2+}, e.g. $CuSO_4$ and CuO

Transition metals can form stable cations with different charges because the successive ionisation energies increase steadily. For example, the extra energy required to remove a third electron from Fe^{2+} is compensated for by the *extra hydration enthalpy* of the 3+ ion compared with the 2+ ion. The third ionisation energy of iron ($Fe^{2+}(g) \rightarrow Fe^{3+}(g) + e^-$) is $+2960\,kJ\,mol^{-1}$, but the hydration energy of the $Fe^{3+}(g)$ ion ($Fe^{3+}(g) + aq \rightarrow Fe^{3+}(aq)$) is $2920\,kJ\,mol^{-1}$ *more exothermic* than the hydration energy of the $Fe^{2+}(g)$ ion.

With calcium, an s-block element, the third ionisation energy is $+4940\,kJ\,mol^{-1}$, which is not compensated for by the extra hydration enthalpy. The third ionisation energy is much higher because the third electron in calcium is removed from an inner $3p$-shell. A transition metal such as chromium can also form oxo-anions, such as CrO_4^{2-}, because it uses all its $4s$- and $3d$-electrons in forming six covalent bonds (4σ and 2π) with oxygen. It accepts six electrons to have the configuration $3d^{10}\,4s^2$.

Exam tip

The second ionisation energy is the energy required to remove the second electron from the 1+ gaseous ion, and is not the energy required to remove two electrons from the gaseous atom.

Exam tip

Scandium forms only Sc^{3+} ions, which have no d-electrons; zinc forms only Zn^{2+} ions, which have ten d-electrons. Therefore, neither is classified as a transition element.

Formation of complex ions

A **complex ion** is formed when a number of ligands (usually four or six) bond to a central metal ion. The bonding is **dative covalent**, with a lone pair of electrons on the ligand forming a bond with empty 3*d*-, 4*s*- or 4*p*-orbitals in the metal ion.

A **ligand** is a species with a lone pair of electrons, which it uses to bond with a central metal ion.

Monodentate ligands

These use one lone pair of electrons to form a dative bond with the metal ion.

- Monodentate ligands can be neutral molecules, such as H_2O and NH_3, or negative ions, such as Cl^- and CN^-.
- All transition metals form hexaaqua hydrated ions, such as $[Cr(H_2O)_6]^{3+}$.
- Zinc, a *d*-block metal, forms the $[Zn(H_2O)_4]^{2+}$ ion.
- In aqua complexes, the bonding between the oxygen atoms and the central metal ion is dative covalent, and the bonding within the water molecule is covalent.
- Fe^{2+} and Fe^{3+} ions form complexes with cyanide ions, $[Fe(CN)_6]^{4-}$ and $[Fe(CN)_6]^{3-}$.
- Cu^{2+} ions form a complex with ammonia — $[Cu(NH_3)_4(H_2O)_2]^{2+}$.
- Chloride ions are much larger than water molecules, so there is a maximum of four Cl^- ions around a transition-metal ion in chloro-complexes. For example, the complexes of copper(I) and copper(II) with chloride ions have the formulae $[CuCl_2]^-$ and $[CuCl_4]^{2-}$.

Bidentate ligands

These are molecules or ions that have two atoms with lone pairs that are both used in forming dative bonds with the metal ion. An example of a neutral bidentate ligand is 1,2-diaminoethane ($NH_2CH_2CH_2NH_2$). This used to be called ethylenediamine and was given the shorthand symbol 'en'. Three 1,2-diaminoethane molecules form six dative bonds with transition-metal ions such as Cu^{2+} (Figure 8):

$$[Cu(H_2O)_6]^{2+}(aq) + 3en(aq) \rightarrow [Cu(en)_3]^{2+}(aq) + 6H_2O(l)$$

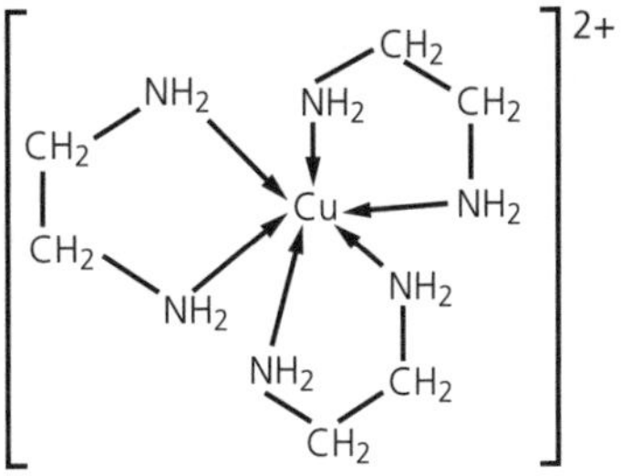

Figure 8

Exam tip

The ring formed between the bidentate ligand and the metal ion must contain five or six atoms, otherwise there would be too much strain on the ring.

Exam tip

The driving force for this ligand-exchange reaction is the gain in entropy as four particles turn into seven particles.

Another example of a bidentate anion is the ethanedioate ion, $^-OOC-COO^-$. This used to be called the oxalate ion and is given the shorthand symbol ox^{2-}.

$$[Cu(H_2O)_6]^{2+}(aq) + 3ox^{2-}(aq) \rightarrow [Cu(ox)_3]^{4-}(aq) + 6H_2O(l)$$

Polydentate ligands

EDTA (the old name was ethylenediaminetetraacetic acid) is a 4− ion that can form six ligands. The stability constants of these complexes are high:

$$[Cu(H_2O)_6]^{2+}(aq) + EDTA^{4-} \rightarrow [Cu(EDTA)]^{2-}(aq) + 6H_2O(l)$$

Haemoglobin contains an iron ion complexed with a pentadentate ligand occupying five sites, with an oxygen atom in the sixth.

Colour of complex ions

The six ligands around the central ion split the five d-orbitals into a group of three of lower energy and a group of two of higher energy. When light is shone into a solution of a complex ion, the ion absorbs light energy and an electron is promoted to the upper of the two split d-levels (Figure 9). As a colour is *removed* from the white light, the solution will have the complementary colour to the colour absorbed. If both red light and yellow light are absorbed, the ion appears blue.

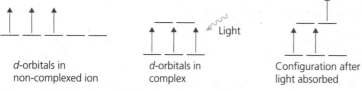

d-orbitals in non-complexed ion   d-orbitals in complex   Configuration after light absorbed

Figure 9 Absorption of light by complex ions

All transition metals form coloured ions in solution as they are surrounded by a number of water molecules that act as ligands.

The colour depends on:

1 the oxidation state of the metal ion
   - $[Cr(H_2O)_6]^{2+}$ is blue; $[Cr(H_2O)_6]^{3+}$ appears green.
   - $[Fe(H_2O)_6]^{2+}$ is green; $[Fe(H_2O)_6]^{3+}$ is amethyst (solutions appear yellow-brown owing to deprotonation of the hydrated ion).

2 the ligand
   - $[Cu(H_2O)_6]^{2+}$ is pale blue; $[Cu(NH_3)_4(H_2O)_2]^{2+}$ is dark blue.

3 the coordination number
   - Hydrated chromium(III) chloride, $[Cr(H_2O)_4Cl_2]^+$ is green and the Cr^{3+} ion has a coordination number of 6, whereas $[CrCl_4]^-$, where it has a coordination number of 4, is yellow.

Lack of colour

- Sc^{3+} and Ti^{4+} ions have no d-electrons. The Cu^+ and Zn^{2+} ions have a full set of 10 d-electrons. The energy levels are split but no d–d transitions are possible. Hence, these ions are colourless.

Knowledge check 20

Explain why diaminomethane, $NH_2CH_2NH_2$ cannot act as a bidentate ligand.

Exam tip

Don't confuse complex ion colours with flame colours, where heat causes an electron to be promoted to an outer orbit and when it falls back coloured light is emitted.

- When hydrated copper(II) sulfate, $CuSO_4.5H_2O$, is heated it changes from a turquoise blue to colourless as the water ligands are driven off. As there are no ligands to cause splitting, the nine d-electrons are all in the same energy level and so the anhydrous salt is colourless.

Shapes of complex ions

All six-coordinate complexes are *octahedral* with 90° bond angles, because there are six pairs of bonding electrons around the metal ion and these six *bond pairs* of electrons repel each other to a position of *maximum separation* (Figure 10).

If the ligands are large, such as Cl^-, the repulsion usually results in only four or sometimes two ligands around the ion. If there are four ligands, the shape is either square-planar, as in the platinum(II) complex $[Pt(NH_3)_2(Cl_2)]$ or tetrahedral, as in the chromium(III) complex $[CrCl_4]^-$. If there are only two ligands, the complex is linear, as in the dichlorocopper(I) anion, $[CuCl_2]^-$.

cis-platin

$[Pt(NH_3)_2(Cl_2)]$ exists as two geometric isomers. Only one, the *cis-* form, is an anticancer drug. It bonds to adjacent guanine molecules in one strand of DNA. This prevents replication of the DNA.

Haemoglobin

In the haemoglobin molecule, the iron ions are complexed with an organic species that can form five bonds to the metal. The sixth position is taken up either by an oxygen molecule (arterial blood) or by a water molecule (venous blood). If a person breathes in carbon monoxide an irreversible ligand exchange reaction occurs with a CO molecule replacing an O_2 molecule. This prevents the haemoglobin from carrying oxygen around the body.

Topic 15B Reactions of transition metal elements

The general properties of transition metal elements are that they:
- have variable valency
- form coloured complexed ions
- undergo ligand exchange reactions
- undergo deprotonation reactions
- are often catalysts

Vanadium

Vanadium exists in several oxidation states (Table 15).

Oxidation state	Formula of ion	Colour
+5	VO_3^-	Colourless
+5	VO_2^+	Yellow
+4	VO^{2+}	Blue
+3	V^{3+}	Green
+2	V^{2+}	Lavender

Table 15 Oxidation states of vanadium

Knowledge check 21

Explain why **a** anhydrous copper(II) sulfate and **b** copper(I) complex ions are both colourless.

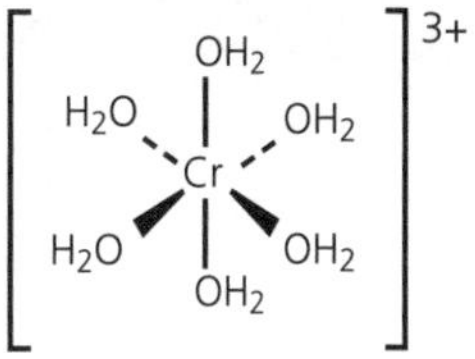

Figure 10 Shape of six-coordinate complex

Exam tip

Do not say that the atoms or the bonds repel. It is the bond pairs of electrons that repel.

Content Guidance

When vanadate(v) ions (VO_3^-) are acidified, an equilibrium is established between two vanadium(v) species.

$$VO_3^-(aq) + 2H^+(aq) \rightleftharpoons VO_2^+(aq) + H_2O(l)$$

Vanadate(v) ions, in acid solution, can be reduced to different extents by different reducing agents.

The $E^\ominus$ values for the different reductions are shown in Table 16.

Oxidation state change	Equation	$E^\ominus/V$
$5 \rightarrow 4$	$VO_2^+ + 2H^+ + e^- \rightleftharpoons VO^{2+} + H_2O$	+1.00
$4 \rightarrow 3$	$VO^{2+} + 2H^+ + e^- \rightleftharpoons V^{3+} + H_2O$	+0.34
$3 \rightarrow 2$	$V^{3+} + e^- \rightleftharpoons V^{2+}$	−0.26

Table 16

- A reducing agent that appears on the *right* of a redox half-equation that has a standard reduction potential value greater than +1.0 V will not reduce vanadium(v) ions.
- A reducing agent that appears on the *right* of a redox half-equation that has a standard reduction potential value between +0.34 and +1.0 V will reduce vanadium(v) to the +4 state but no further.
- A reducing agent that appears on the *right* of a redox half-equation that has a standard reduction potential value between −0.26 V and +0.34 V will reduce vanadium(v) to the +3 state but not to +2.
- A reducing agent that appears on the *right* of a redox half-equation that has a standard reduction potential more negative than −0.26 V will reduce vanadium(v) from the +5 to the +2 state.

Worked example

How far will a solution of Fe^{2+} ions reduce a solution of VO_2^+?

$$Fe^{3+} + e^- \rightleftharpoons Fe^{2+} \quad E^\ominus = +0.77\,V$$

Answer

The figure +0.77 V is between +0.34 V and +1.0 V, and so Fe^{2+} should reduce vanadium(v) to vanadium(iv) and no further. To prove this, reverse the Fe^{3+}/Fe^{2+} equation and add it to the one for vanadium changing oxidation state from +5 to +4.

$$Fe^{2+} + VO_2^+ + 2H^+ \rightleftharpoons Fe^{3+} + VO^{2+} + H_2O$$

$$E^\ominus_{cell} = +1.0 + (-0.77) = +0.23\,V$$

As $E^\ominus_{cell}$ is positive, the reaction will occur. Therefore, Fe^{2+} will reduce vanadium +5 to the +4 state. However, reversing the Fe^{3+}/Fe^{2+} equation and adding it to the one for vanadium +4 to +3 gives:

$$Fe^{2+} + VO^{2+} + 2H^+ \rightleftharpoons Fe^{3+} + V^{3+} + H_2O$$

$$E^\ominus_{cell} = +0.34 + (-0.77) = -0.43\,V$$

$E^\ominus_{cell}$ is negative, so the reaction will not occur. The Fe^{2+} ions will not reduce the vanadium to the +3 state. This can be tested in the laboratory by adding excess iron(iii) sulfate solution to acidified ammonium vanadate(v). The solution goes from yellow to blue, showing that the vanadium is now in the +4 state.

Exam tip

As both vanadium species are in the +5 state, this is *not* a redox reaction.

Exam tip

This can also be shown using the anticlockwise rule. The electrode system with the reducing agent (on the right) must be above (less positive than) the electrode system for the vanadium species in the table in the *Data Booklet*.

Knowledge check 22

Given $S(s) + 2H^+(aq) + 2e^- \rightleftharpoons H_2S(g)$ $E^\ominus = +0.14\,V$

Use the data in Table 16 to predict the oxidation state of vanadium to which hydrogen sulfide will reduce vanadate(v) ions in acid solution.

Chromium

Chromium can exist in the +6 state, as in compounds such as potassium dichromate(VI) ($K_2Cr_2O_7$), which is orange, and potassium chromate(VI) (K_2CrO_4), which is yellow. It also forms many compounds in the +3 state, such as chromium(III) sulfate ($Cr_2(SO_4)_3$) and many complexes of Cr^{3+} ions. Most chromium(III) compounds are green.

Some chromium(II) compounds also exist, but most of these are unstable. The chromium(II) complex with ethanoate ions is stable. It is an unusual complex, in that it contains two Cr^{2+} ions.

The standard reduction potentials are listed below:

$$\tfrac{1}{2}Cr_2O_7^{2-} + 7H^+ + 3e^- \rightleftharpoons Cr^{3+} + \tfrac{7}{2}H_2O \qquad E^\ominus = +1.33\,V$$

$$Cr^{3+} + e^- \rightleftharpoons Cr^{2+} \qquad\qquad\qquad E^\ominus = -0.41\,V$$

Chromium(VI)

There are two ionic species in which the oxidation state of chromium is +6:

- dichromate(VI) ions, $Cr_2O_7^{2-}(aq)$, which are orange
- chromate(VI) ions, $CrO_4^{2-}(aq)$, which are yellow

They are in equilibrium:

$$2CrO_4^{2-}(aq) + 2H^+(aq) \rightleftharpoons Cr_2O_7^{2-}(aq)$$

Reduction of chromium compounds

A reducing agent that appears on the *right* of a redox half-equation and which has a standard reduction potential value of less than +1.33 V will reduce dichromate(VI) ions in acid solution under standard conditions.

A reducing agent that appears on the *right* of a redox half-equation and which has a standard reduction potential value that is *more* negative than −0.41 V will reduce Cr^{3+} to Cr^{2+}.

Worked example

Will zinc metal, in acid solution, reduce $Cr_2O_7^{2-}$ ions to Cr^{2+} ions?

$$Zn^{2+} + 2e^- \rightleftharpoons Zn \quad E^\ominus = -0.76\,V$$

Answer

For +6 to +3 (using method 1 as on page 43):

$$E^\ominus_{cell} = E^\ominus(Cr_2O_7^{2-}/Cr^{3+}) - E^\ominus(Zn^{2+}/Zn) = +1.33 - (-0.76) = +2.09\,V$$

Or using the anticlockwise rule as on page 44:

$$Zn^{2+} + 2e^- \rightleftharpoons Zn \quad E^\ominus = -0.76\,V$$

$$\tfrac{1}{2}Cr_2O_7^{2-} + 7H^+ + 3e^- \rightleftharpoons Cr^{3+} + \tfrac{7}{2}H_2O \qquad E^\ominus = +1.33\,V$$

$$E^\ominus_{cell} = +1.33 - (-0.76) = +2.09\,V$$

➡

Knowledge check 23

State and explain the observations when aqueous alkali is added to a solution of potassium dichromate(VI).

This is a positive number, so zinc metal will reduce acidified dichromate ions under standard conditions.

For +3 to +2:

$$E^{\ominus}_{cell} = E^{\ominus}(Cr^{3+}/Cr^{2+}) - E^{\ominus}(Zn^{2+}/Zn) = -0.41 - (-0.76) = +0.35\,V$$

This is also a positive value, so zinc in acid solution will reduce Cr^{3+} ions to Cr^{2+} ions.

Oxidation of chromium compounds

Chromium(III) is oxidised to chromium(VI) in alkaline solutions:

$$CrO_4^{2-} + 4H_2O + 3e^- \rightleftharpoons Cr(OH)_3 + 5OH^- \qquad E^{\ominus} = -0.13\,V$$

An oxidising agent that appears on the *left* of a redox half-equation and which has a standard reduction potential that is more positive (or less negative) than $-0.13\,V$ will oxidise chromium(III) compounds to chromate(VI) in alkaline solution.

Worked example

Will hydrogen peroxide, H_2O_2, in alkaline solution, oxidise chromium(III) hydroxide to chromate(VI) ions?

$$H_2O_2 + 2e^- \rightleftharpoons 2OH^- \qquad E^{\ominus} = +1.24\,V$$

Answer

$$E^{\ominus}_{cell} = E^{\ominus}(H_2O_2/OH^-) - E^{\ominus}(CrO_4^{2-}/Cr(OH)_3) = +1.24 - (-0.13) = +1.37\,V$$

This is a positive number so hydrogen peroxide, in alkaline solution, will oxidise chromium(III) to chromate(VI).

Deprotonation

With water

All aqua ions of transition metals are to some extent deprotonated by water. The extent depends on the charge density of the metal ion. The greater the charge and the smaller the radius, the more the aqua ion is deprotonated.

Solutions of hydrated chromium(III) ions are acidic, owing to the production of H_3O^+ ions:

$$[Cr(H_2O)_6]^{3+}(aq) + H_2O(l) \rightleftharpoons [Cr(H_2O)_5OH]^{2+}(aq) + H_3O^+(aq)$$

With aqueous sodium hydroxide

All the hydrated ions are deprotonated by $OH^-(aq)$ ions, forming the neutral hydrated hydroxide as a precipitate:

$$[M(H_2O)_6]^{2+}(aq) + 2OH^-(aq) \rightarrow [M(H_2O)_4(OH)_2](s) + 2H_2O(l)$$

$$[M(H_2O)_6]^{3+}(aq) + 3OH^-(aq) \rightarrow [M(H_2O)_3(OH)_3](s) + 3H_2O(l)$$

where M stands for any *d*-block metal cation.

Knowledge check 24

Will bromide ions reduce acidified dichromate ions to Cr^{3+} ions but no further under standard conditions?

$$½Br_2 + e^- \rightleftharpoons Br^-$$
$$E^{\ominus} = +1.07\,V$$

Knowledge check 25

Will Fe^{3+} ions oxidise Cr^{3+} to $Cr_2O_7^{2-}$ ions in acid solution? Use the *Data Booklet* for $E^{\ominus}$ values.

Exam tip

The greater the charge density of the metal ion, the more acidic is the solution, so a solution of chromium(III) ions is more acidic than one of copper(II) ions.

Note that the number of hydroxide ions on the left-hand side of the equation and the number of water molecules on the right-hand side are both equal to the charge on the d-block metal ion.

The colour changes are shown in Table 17.

Complex	Colour of complex	Colour of precipitate	Comment
$[Cr(H_2O)_6]^{3+}$	Green solution	Green precipitate	—
$[Mn(H_2O)_6]^{2+}$	Pale pink solution	Sandy precipitate that darkens in air	Manganese is oxidised to the +4 state
$[Fe(H_2O)_6]^{2+}$	Pale green solution	Green precipitate that goes brown in air	Iron is oxidised to the +3 state
$[Fe(H_2O)_6]^{3+}$	Yellow solution	Red-brown precipitate	—
$[Co(H_2O)_6]^{2+}$	Pink solution	Blue precipitate that goes red on standing	The hydrated hydroxide loses water
$[Ni(H_2O)_6]^{2+}$	Green solution	Pale green precipitate	—
$[Cu(H_2O)_6]^{2+}$	Blue solution	Blue precipitate	—
$[Zn(H_2O)_6]^{2+}$	Colourless solution	White precipitate	—

Table 17 Colour changes with aqueous sodium hydroxide

Reactions with excess aqueous sodium hydroxide

Amphoteric hydroxides react with excess sodium hydroxide to form a solution.

An amphoteric oxide or hydroxide is one that reacts with acids to form hydrated cations:

$$[Cr(H_2O)_3(OH)_3](s) + 3H^+(aq) \rightarrow [Cr(H_2O)_6]^{3+}(aq) + 3H_2O(l)$$

and also with bases to form a hydroxyanion.

On adding excess sodium hydroxide, they are further deprotonated — for example, chromium and zinc hydroxides:

$$[Cr(H_2O)_3(OH)_3](s) + 3OH^-(aq) \rightarrow [Cr(OH)_6]^{3-}(aq) + 3H_2O(l)$$
dark green

$$[Zn(H_2O)_2(OH)_2](s) + 2OH^-(aq) \rightarrow [Zn(OH)_4]^{2-}(aq) + 2H_2O(l)$$
colourless

Reactions with aqueous ammonia

Ammonia is a base and a ligand. When ammonia solution is added to a solution of a transition-metal salt, the hydrated metal ion is deprotonated, as with sodium hydroxide solution, giving the same coloured precipitates. For example:

$$[Fe(H_2O)_6]^{2+}(aq) + 2NH_3(aq) \rightarrow [Fe(H_2O)_4(OH)_2](s) + 2NH_4^+(aq)$$

$$[Fe(H_2O)_6]^{3+}(aq) + 3NH_3(aq) \rightarrow [Fe(H_2O)_3(OH)_3](s) + 3NH_4^+(aq)$$

However, excess ammonia results in **ligand exchange** with nickel, copper and zinc.

$$[Ni(H_2O)_4(OH)_2](s) + 4NH_3(aq) \rightarrow [Ni(NH_3)_4(H_2O)_2]^{2+}(aq) + 2OH^-(aq) + 2H_2O(l)$$
pale blue

$$[Cu(H_2O)_4(OH)_2](s) + 4NH_3(aq) \rightarrow [Cu(NH_3)_4(H_2O)_2]^{2+}(aq) + 2OH^-(aq) + 2H_2O(l)$$
dark blue

$$[Zn(H_2O)_2(OH)_2](s) + 4NH_3(aq) \rightarrow [Zn(NH_3)_4]^{2+}(aq) + 2OH^-(aq) + 2H_2O(l)$$
colourless

Ligand exchange

These are reactions in which one ligand either totally or partially replaces the ligand in a complex. For example, when an EDTA solution is added to a solution of the hexaaquachromium(III) ion, the EDTA complex is formed in a ligand-exchange reaction:

$$[Cr(H_2O)_6]^{3+} + EDTA^{4-} \rightarrow [Cr(EDTA)]^- + 6H_2O$$

If cyanide ions are added to hexaaqua iron(III) ions, ligand exchange takes place:

$$[Fe(H_2O)_6]^{3+} + 6CN^- \rightarrow [Fe(CN)_6]^{3-} + 6H_2O$$

If excess ammonia solution is added to precipitates of the hydroxides of copper(II), zinc, nickel, cobalt or silver, ligand exchange takes place, forming a solution of the ammonia complex of the *d*-block ion.

$$[Cu(H_2O)_4(OH)_2] + 4NH_3 \rightarrow [Cu(NH_3)_4(H_2O)_2]^{2+} + 2OH^- + 2H_2O$$

$$[Zn(H_2O)_2(OH)_2] + 4NH_3 \rightarrow [Zn(NH_3)_4]^{2+} + 2OH^- + 2H_2O$$

When concentrated hydrochloric acid is added to a pale blue solution of hydrated copper(II) ions, the solution first turns green and finally goes yellow as a ligand-exchange reaction takes place.

$$[Cu(H_2O)_6]^{2+} + 4Cl^- \rightarrow [CuCl_4]^{2-} + 6H_2O$$
pale blue   yellow

The green colour is produced when the two copper complex ions are in similar concentrations.

When concentrated hydrochloric acid is added to pink aqueous cobalt(II) sulfate the solution turns blue as $[CoCl_4]^{2-}$ complex ions are formed.

$$[Co(H_2O)_6]^{2+}(aq) + 4Cl^-(aq) \rightarrow [CoCl_4]^{2-}(aq) + 6H_2O(l)$$

In both this, and the reaction above, the coordination number changes from 6 to 4. This is because there would be too much repulsion in fitting six chloride ions around the metal cation as Cl^- is larger than H_2O.

Catalytic activity

Transition metals and their compounds are often excellent catalysts.

Heterogeneous catalysis

This is when the catalyst and the reactants are in different phases. The metals use their *d*-orbitals to provide active sites on their surfaces to which reactants bond. For example, nickel is the catalyst for the addition of hydrogen to alkene. Iron is used in the Haber process, catalysing the reaction:

$$N_2 + 3H_2 \rightleftharpoons 2NH_3$$

Exam tip

Silver chloride and silver bromide are insoluble but dissolve in ammonia solution to form $[Ag(NH_3)_2]^+$ ions. This is the basis of the test for halides.

Exam tip

The driving force for this reaction is the increase in entropy. Two particles become seven.

Exam tip

The colour of the copper compound depends on the ligand. This is the easiest way to see if ligand exchange has taken place.

Compounds of transition metals can change oxidation state and this is made use of in industrial processes. Vanadium(v) oxide is the catalyst in the manufacture of sulfuric acid and catalyses the reaction:

$$2SO_2 + O_2 \rightleftharpoons 2SO_3$$

The compound can do this because of the variable valency of vanadium. The mechanism is:

- Step 1: $2SO_2 + 2V_2O_5 \rightarrow 2SO_3 + 4VO_2$
- Step 2: $4VO_2 + O_2 \rightarrow 2V_2O_5$
- Overall: $2SO_2 + O_2 \rightarrow 2SO_3$

Cars are fitted with catalytic converters. These are platinum coated on the surface of an inert solid. It works in three steps:

1 Adsorption onto the active sites on the surface of the catalyst. This weakens the bonds in the reactant.

2 Reaction of adsorbed reactants to adsorbed products.

3 Desorption of product molecules from the catalyst's surface forming new active sites ready to adsorb more reactant molecules.

Homogeneous catalysis

This is where the catalyst and the reactants are in the same phase either as gases or when dissolved in water.

Fe^{3+} ions catalyse the oxidation of iodide ions by persulfate ions ($S_2O_8^{2-}$). The mechanism is:

- Step 1: $2Fe^{3+}(aq) + 2I^-(aq) \rightarrow 2Fe^{2+}(aq) + I_2(s)$
- Step 2: $2Fe^{2+}(aq) + S_2O_8^{2-}(aq) \rightarrow 2Fe^{3+}(aq) + 2SO_4^{2-}(aq)$
- Overall: $2I^-(aq) + S_2O_8^{2-}(aq) \rightarrow I_2(s) + 2SO_4^{2-}(aq)$

The redox reaction where MnO_4^- ions are reduced by $C_2O_4^{2-}$ ions is catalysed by Mn^{2+} ions. These are produced by the reduction of MnO_4^- ions and so the reaction is an example of *autocatalysis*.

Core practical 12

The preparation of a transition metal complex

A solution of potassium dichromate(vi) is placed in a flask with some zinc and a mixture of 50% concentrated hydrochloric acid and water is added. A delivery tube is fitted through a screw seal with the bottom end below the surface of the acidified potassium dichromate(vi) solution and the other end below the surface of a solution of sodium ethanoate.

At first the seal at the top is left completely loose in order to let out the hydrogen. The orange solution turns green as Cr^{3+} ions are formed, and then blue as these are reduced to Cr^{2+} ions. At this stage, the cap of the seal is screwed shut and the pressure of hydrogen forces the solution containing Cr^{2+} ions out into the test tube.

The hydrated chromium(ii) ions undergo ligand exchange, forming a precipitate of the red chromium(ii) ethanoate complex.

$$2[Cr(H_2O)_6]^{2+}(aq) + 4CH_3COO^-(aq) \rightarrow [Cr_2(CH_3COO)_4(H_2O)_2](s) + 10H_2O(l)$$

Exam tip

In core practical 12, air must be excluded from the apparatus or else the Cr^{2+} ions formed will immediately be oxidised by oxygen to Cr^{3+} ions.

Content Guidance

Summary

After studying this topic, you should be able to:

- define transition metals and list their typical properties
- write the electron configurations of transition metals and their positive ions
- explain the shape of complex ions and why most of them are coloured
- understand the difference between monodentate, bidentate and polydentate ligands
- explain redox reactions of transition metal ions in terms of electrode potentials
- describe the colours and oxidation states of vanadium
- use $E^\ominus$ values to predict the feasibility of redox reactions involving chromium in different oxidation states
- give examples of ligand exchange and deprotonation reactions
- understand what is the driving force for the substitution reactions involving bidentate and polydentate ligands
- understand why transition metals and their compounds can be catalysts
- describe the preparation of a transition metal complex

Questions & Answers

This section contains multiple-choice and structured questions similar to those you can expect to find in A-level paper 1 and in parts of questions in synoptic A-level paper 3. The questions given here are not balanced in terms of types of question or level of demand — they are not intended to typify real papers, only the sorts of questions that could be asked.

In the examinations, the answers are written on the question paper. Here, the questions are not shown in examination paper format.

The answers given are those that would be expected from a top-grade student. They are not 'model answers' to be regurgitated without understanding. In answers that require more extended writing, it is usually the ideas that count rather than the form of words used. The principle is that correct and relevant chemistry scores.

Comments

Comments on the questions are indicated by the icon ⓔ. They offer tips on what you need to do to get full marks. Responses to questions may have a comment, preceded by the icon ⓔ. The comments may explain the correct answer, point out common errors made by students who produce work of C-grade or lower, or contain additional useful advice.

The exam papers

A-level paper 1 examines topics 1–15, lasts 1 hour 45 minutes and is worth 90 marks. Paper 3 is a synoptic paper that examines all the topics on the specification and lasts 2 hours 30 minutes with a maximum of 120 marks. Both papers may include multiple-choice, short open, open-response, calculations and extended writing questions. Note that over all three A-level papers, a minimum of 20% of the marks will be awarded for mathematics at level 2 or above. The vast majority of practical based questions will be on paper 3 and are covered in the fourth student guide of this series.

Assessment objectives

The AS and A-level exams have three assessment objectives (AOs). The percentages of each are very similar in the AS and the A-level papers, with a heavier weighting of AO3 in the A-level paper.

AO1 is 'knowledge and understanding of scientific ideas, processes and procedures'. This makes up about 30% of the total A-level exam. You should be able to:

- recognise, recall and show understanding of scientific knowledge
- select, organise and present information clearly and logically, using specialist vocabulary where appropriate

AO2 is 'application of knowledge and understanding of scientific ideas, processes and procedures'. This makes up about 40% of the total A-level exam. You should be able to:

- analyse and evaluate scientific knowledge and processes
- apply scientific knowledge and processes to unfamiliar situations
- assess the validity, reliability and credibility of scientific information

AO3 is 'the analysis, interpretation and evaluation of scientific ideas and data'. This makes up about 25% of the total A-level exam. You should be able to make scientific judgements, reach conclusions and evaluate and improve described practical procedures.

Command terms

Examiners use certain words that require you to respond in a specific way. You must distinguish between these terms and understand exactly what each requires you to do.

- **Define** — give a simple definition without any explanation.
- **Identify** — give the name or formula of the substance.
- **State** — no explanation is required (nor should you give one).
- **Deduce** — use the information supplied in the question to work out your answer.
- **Suggest** — use your knowledge and understanding of similar substances or those with the same functional groups to work out the answer.
- **Compare** — make a statement about *both* substances being compared.
- **Explain** — use chemical theories or principles to say why a particular property is as it is.
- **Predict** — say what you think will happen on the basis of the principles that you have learned.

Revision

Start your revision in plenty of time. Make a list of the things that you need to do, emphasising the things that you find most difficult, and draw up a detailed revision plan. Work back from the examination date, ideally leaving an entire week free from fresh revision before that date. Be realistic in your revision plan and then add 25% to the timings because everything takes longer than you think.

When revising:

- make a note of difficulties and ask your teacher about them — if you do not make such notes, you will forget to ask
- make use of past papers, but remember that these have been written to a different specification
- revise ideas, rather than forms of words — you are after understanding
- remember that scholarship requires time to be spent on the work

When you use the example questions in this book, make a determined effort to answer them before looking up the answers and comments. Remember that the answers given here are not intended as model answers to be learnt parrot-fashion. They are answers designed to illuminate chemical ideas and understanding.

The exam papers

- Read the question. Questions usually change from one examination to the next. A question that looks the same, at a cursory glance, as one that you have seen before usually has significant differences when read carefully. Needless to say, students do not receive credit for writing answers to their own questions.

- Be aware of the number of marks available for a question. That is an excellent pointer to the number of things you need to say.

- Do not repeat the question in your answer. The danger is that you fill up the space available and think that you have answered the question, when in reality some or maybe all of the real points have been ignored.

- The name of a 'rule' is not an explanation for a chemical phenomenon. For example, in equilibrium, a common answer to a question on the effect of changing pressure on an equilibrium system is 'Because of Le Chatelier's principle...'. That is simply a name for a rule — it does not explain anything.

Multiple-choice questions

Answers to multiple-choice questions are machine-marked. Multiple-choice questions need to be read carefully; it is important not to jump to a conclusion about the answer too quickly. You need to be aware that one of the options might be a 'distracter'. An example of this might be in a question having a numerical answer of, say, $-600\,\text{kJ}\,\text{mol}^{-1}$; a likely distracter would be $+600\,\text{kJ}\,\text{mol}^{-1}$.

Some questions require you to think on paper — there is no demand that multiple-choice questions be carried out in your head. Space is provided on the question paper for rough working. It will not be marked, so do not write anything that matters in this space because no-one will see it.

For each of the questions there are four suggested answers, A, B, C and D. You select the best answer by putting a cross in the box beside the letter of your choice. If you change your mind you should put a line through the box and then indicate your alternative choice. Making more than one choice does not earn any marks. Note that this format is not used in the multiple-choice questions in this book.

Each test has at least 10 multiple-choice questions, which are embedded in longer questions.

Online marking

It is important that you have some understanding of how examinations are marked, because to some extent it affects how you answer them. Your examination technique partly concerns chemistry and partly must be geared to how the examinations are dealt with physically. You have to pay attention to the layout of what you write. Because all your scripts are scanned and marked online, there are certain things you must do to ensure that all your work is seen and marked. This is covered below.

As the examiner reads your answer, decisions have to be made — is this answer worth the mark or not? Your job is to give the *clearest possible answer* to the question asked, in such a way that your chemical understanding is made obvious to the examiner. In particular, you must not expect the examiner to guess what is in your head; you can be judged only by what you write.

It is especially important that you *think before you write*. You will have a space on the question paper that the examiner has judged to be a reasonable amount for the answer. Because of differing handwriting sizes, false starts and crossings-out and because some students have a tendency to repeat the question in the answer space, the space is never exactly right for all students.

Edexcel exam scripts are marked online, so few examiners will handle a real, original script. The process is as follows:

- When the paper is set it is divided up into items, often, but not necessarily, a single part of a question. These items are also called clips.
- The items are set up so that they display on-screen, with check-boxes for the score and various buttons to allow the score to be submitted or for the item to be processed in some other way.
- After you have written your paper it is scanned; from that point all the handling of your paper is electronic. Your answers are tagged with an identity number.
- It is impossible for an examiner to identify a centre or a student from any of the information supplied.

Common pitfalls

The following list of potential pitfalls to avoid is particularly important:

- Do not write in any colour other than black — this is an exam board regulation. The scans are entirely black-and-white, so any colour used simply comes out black — unless you write in red, in which case it does not come out at all. The scanner cannot see red (or pink or orange) writing. So if, for example, you want to highlight different areas under a graph, or distinguish lines on a graph, you must use a different sort of shading rather than a different colour.
- Do not use small writing. Because the answer appears on a screen, the definition is slightly degraded. In particular, small numbers used for powers of 10 can be difficult to see. The original script is always available but it can take a long time to get hold of it.
- Do not write in pencil. Faint writing does not scan well.
- Do not write outside the space provided without saying, within that space, where the remainder of the answer can be found. Examiners only have access to a given item; they cannot see any other part of your script. So if you carry on your answer elsewhere but do not tell the examiner within the clip that it exists, it will not be seen. Although the examiner cannot mark the out-of-clip work, the paper will be referred to the Principal Examiner responsible for the paper.
- Do not use asterisks or arrows as a means of directing examiners where to look for out-of-clip items. Tell them in words. Students use asterisks for all sorts of things and examiners cannot be expected to guess what they mean.
- Do not write across the centre-fold of the paper from the left-hand to the right-hand page. A strip about 8 mm wide is lost when the papers are guillotined for scanning.
- Do not repeat the question in your answer. If you have a question such as 'Define the first ionisation energy of calcium', the answer is 'The energy change per mole for the formation of unipositive ions from isolated calcium atoms in the gas phase'; or, using the equation, 'The energy change per mole for $Ca(g) \rightarrow Ca^+(g) + e^-$'. Do not start by writing 'The first ionisation energy for calcium is defined as...' because by then you will have taken up most of the space available for the answer. Examiners know what the question is — they can see it on the paper.

■ Structured and multiple-choice questions

Question 1

(a) Consider the equilibrium reaction:

$A(g) + B(g) \rightleftharpoons C(g)$ $\Delta H = +23\,kJ\,mol^{-1}$

Which of the following statements is true? (1 mark)

A An increase in pressure will drive the position of equilibrium to the right and hence will increase the value of the equilibrium constant, K.

B An increase in pressure will increase the value of the equilibrium constant, K, and hence drive the position of equilibrium to the right.

C An increase in temperature will drive the position of equilibrium to the right and hence will increase the value of the equilibrium constant, K.

D An increase in temperature will increase the value of the equilibrium constant, K, and hence drive the position of equilibrium to the right.

(b) Hydrogen can be made by reacting steam with coke at high temperature:

$C(s) + H_2O(g) \rightleftharpoons CO(g) + H_2(g)$ $\Delta H = +131\,kJ\,mol^{-1}$

At a pressure of 1.2 atm, 90% of the steam was converted into carbon monoxide and hydrogen.

(i) Write the expression for K_p (1 mark)

ⓔ K_p expressions do not have square brackets.

(ii) Calculate the value of K_p at this temperature to two significant figures. Include the units in your answer. (6 marks)

ⓔ This is best answered in a table. The marks are for equilibrium moles, mole fractions, partial pressures, the value of K_p and its units. Do not reduce intermediate values to 2 significant figures, only the final answer.

(c) For the reaction in (b), state and explain the effect on the equilibrium constant, K_p, and *hence* on the equilibrium yield of hydrogen of:

(i) increasing the pressure to 2.4 atm (4 marks)

ⓔ First, state whether or not K changes. Second, investigate whether the partial pressure term and K are still equal and, if not, what must then happen. Finish your answer with a statement about the equilibrium yield of hydrogen.

(ii) increasing the temperature (3 marks)

ⓔ Consider the change, if any, to the value of K for this endothermic reaction. Then continue as in your answer to (i).

(d) Suggest why a temperature of 2000°C is not used. (1 mark)

ⓔ The choice of temperature is determined by yield, rate and cost.

Total: 17 marks

(e) The Haber process for the manufacture of ammonia utilises the exothermic reaction between nitrogen and hydrogen. It is carried out at 400°C at a pressure of 200 atm in the presence of a heterogeneous catalyst of iron.

$$N_2(g) + 3H_2(g) \rightleftharpoons 2NH_3(g) \quad \Delta H = -92\,kJ\,mol^{-1}$$

Which of the following statements is **not** true about this process?   (1 mark)

A   High pressure is used to increase the equilibrium yield.

B   High pressure is used to speed up the rate of reaction.

C   The value of the equilibrium constant is small at 400°C and is even smaller at 500°C.

D   A temperature of 400°C and an iron catalyst are chosen because these conditions result in a reasonable yield at a fast rate.

Student answer
(a) D ✓

e A change in pressure will not alter the value of the equilibrium constant, K, so options **A** and **B** cannot be correct. The logic in option **C** is wrong, but is correct in option **D**. For any endothermic reaction, an increase in temperature will cause an increase in the value of K, which becomes bigger than the equilibrium term [products]/[reactants]. This term has to get larger, so that it once again equals the value of K. To do this, the value of the upper line must increase and that of the bottom line decrease. This happens as the position of equilibrium shifts to the right, so the answer is **D**.

(b) (i) $K_p = \dfrac{p(CO)p(H_2)}{p(H_2O)}$ ✓

e A value for carbon (coke) is not included because it is a solid.

(ii)

	$H_2O(g)$	$CO(g)$	$H_2(g)$	
Initial moles	1	0	0	
Equilibrium moles	$1 - 0.10 = 0.90$	0.10	0.10	✓
Mole fraction	$\dfrac{0.90}{1.10} = 0.818$	$\dfrac{0.10}{1.10} = 0.0909$	$\dfrac{0.10}{1.10} = 0.0909$	✓
Partial pressure atm	$0.818 \times 1.2 = 0.982$	$0.0909 \times 1.2 = 0.109$	$0.0909 \times 1.2 = 0.109$	✓

$$K_p = \frac{p(CO)p(H_2)}{p(H_2O)} = \frac{0.109\,atm \times 0.109\,atm}{0.982\,atm} = 0.012 \checkmark\ atm\ (2\ s.f.)\ \checkmark$$

e The route is:

% → equilibrium moles → mole fraction → partial pressure → value for
K_p → units of K_p

(c) (i) Increasing the pressure has no effect on the value of K_p ✓; only temperature change affects this. However, the partial pressure term increases by a factor of two and so is no longer equal to the unaltered K_p ✓. The system is now not at equilibrium and reacts by moving to the left until the partial pressure term once again equals K_p ✓. Thus the equilibrium yield of hydrogen is decreased ✓.

(ii) Increasing the temperature will cause the value of K_p to increase, as the reaction is endothermic ✓. The partial pressure term is now smaller than K_p, so the system is not in equilibrium ✓ and reacts to make more hydrogen and carbon monoxide until the partial pressure term equals the new larger value of K_p ✓.

(d) This temperature would result in greater conversion to hydrogen, but would be uneconomic as water is cheap and such a high temperature is expensive to maintain ✓.

(e) B ✓

ⓔ This question is best answered by putting true or false beside each option:

- Option A: True — there are fewer gas moles on the right, so an increase in pressure will drive the equilibrium to the right and so increase the yield.
- Option B: False — the rate of a reaction involving a solid metallic catalyst at a given temperature is not altered by pressure. It depends solely on the number of active sites on the surface of the catalyst.
- Option C: True — the value of K is small at 400°C (which is why it is economic to use high pressure). Since the reaction is exothermic, the value of K will decrease as the temperature rises.
- Option D: True — a catalyst allows the reaction to proceed at a fast rate at a lower temperature. This results in a higher yield than if a higher temperature were used.

Options A, C and D are all true. Option B is false and is the correct answer to this negative question.

Question 2

Nitrous acid, HNO_2, is a weak acid.

(a) (i) Write the expression for K_a. (1 mark)

(ii) The best statement about a weak acid is: (1 mark)

    A it is partially ionised in aqueous solution

    B it is only slightly ionised in aqueous solution

    C its pH will be 2 units more than that of a strong acid of equal concentration

    D its pH will be 2 units less than that of a strong acid of equal concentration

(b) The pK_a of nitrous acid at 25°C is 3.34. The pH of a solution was found to be 2.65. Calculate the concentration of the solution in $mol\,dm^{-3}$.　　　　　(3 marks)

ⓔ You will need to use your answer to (a)(i) to calculate this.

(c) The concentration could also be found by titrating the nitrous acid solution against a standard solution of sodium hydroxide.

The best indicator for this titration would be:　　　　　(1 mark)

	Indicator	pK_{ind}
A	Bromophenol blue	4.0
B	Bromothymol blue	7.0
C	Thymol blue	8.9
D	Alizarin yellow R	12.5

ⓔ Decide on the approximate pH at the end point in this weak acid/strong base titration.　　　　　Total: 6 marks

Student answer

(a) (i) $K_a = \dfrac{[H^+] \times [NO_2^-]}{[HNO_2]}$ or $K_a = \dfrac{[H_3O^+] \times [NO_2^-]}{[HNO_2]}$ ✓

(ii) B ✓

ⓔ The ionisation is normally less than 10% so slightly is a better answer than partially. The rule of 2 is only approximate so C is not the best answer and D is wrong as the pH will be more not less than that of a strong acid.

(b) $pK_a = 3.34$　so $K_a = 10^{-3.34} = 4.57 \times 10^{-4}$ ✓

$pH = 2.65$　so $[H^+] = 10^{-2.65} = 0.002239$ ✓

$[HNO_2] = \dfrac{[H^+]^2}{K_a} = 0.0110\,(mol\,dm^{-3})$ ✓

(c) C ✓

ⓔ The end point for a weak acid/strong base titration will be above 7 and below 11.

Question 3

Propanoic acid, CH_3CH_2COOH, is a weak acid that reacts reversibly with water:
$$CH_3CH_2COOH(aq) + H_2O(l) \rightleftharpoons H_3O^+(aq) + CH_3CH_2COO^-(aq)$$

When an aqueous solution of propanoic acid is mixed with solid sodium propanoate, a buffer solution is formed.

(a) Write the expression for the acid dissociation constant, K_a, of propanoic acid.　　　　　(1 mark)

ⓔ Think whether or not this expression contains $[H_2O]$.

(b) (i) Define a buffer solution.　　　　　(2 marks)

ⓔ Be careful. Does the pH of a buffer ever change?

(ii) A buffer solution of a weak acid and its salt has a pH of 5.00. When it is diluted ten-fold, its pH:

  A stays the same

  B alters by a small amount only, because a buffer resists changes in pH

  C falls to 4.00

  D increases to 6.00   (1 mark)

(iii) Calculate the mass of sodium propanoate that has to be added to $100\,cm^3$ of a $0.600\,mol\,dm^{-3}$ solution of propanoic acid to make a buffer solution of pH 5.06 at 25°C. (The acid dissociation constant of propanoic acid $= 1.31 \times 10^{-5}\,mol\,dm^{-3}$ at 25°C; molar mass of sodium propanoate $= 96.0\,g\,mol^{-1}$.)   (4 marks)

ⓔ Use the expression for K_a and insert values for K_a, [acid] and $[H^+]$. Solve for [salt], then work out the moles and hence the mass.

(iv) Explain why, when a small amount of OH^- ions is added, a solution of propanoic acid alone is not a buffer solution.   (3 marks)

ⓔ This is a discriminating question. You should write an equation for the addition of OH^- ions and discuss the effect on [acid] and [conjugate base] of this reaction.   Total: 11 marks

---

Student answer

(a) $K_a = \dfrac{[H_3O^+][CH_3CH_2COO^-]}{[CH_3CH_2COOH]}$ ✓

(b) (i) A buffer solution is one that maintains a nearly constant pH ✓, when small amounts of acid or alkali are added ✓.

  (ii) A ✓.

ⓔ $[H^+] = K_a[acid]/[salt]$, so the pH of a buffer depends on the ratio of [acid] to [salt]. This ratio is not altered by dilution, so the pH stays the same. Option B shows a misunderstanding of the term 'buffer'. A buffer resists change in pH when a small amount of *acid* or *alkali* is added. Options C and D would be true if the ratio of [acid] to [salt] were to increase or decrease by a factor of 10.

(iii) $[H_3O^+] = 10^{-pH} = 10^{-5.06} = 8.710 \times 10^{-6}\,mol\,dm^{-3}$ ✓

$[CH_3CH_2COO^-] = \dfrac{K_a[CH_3CH_2COOH]}{[H_3O^+]} = \dfrac{1.31 \times 10^{-5} \times 0.600}{8.710 \times 10^{-6}} = 0.9024\,mol\,dm^{-3}$ ✓

moles of salt $=$ concentration $\times$ volume $= 0.9024\,mol\,dm^{-3} \times 0.100\,dm^3$
$= 0.09024\,mol$ ✓

mass of salt needed $=$ moles $\times$ molar mass $= 0.09024 \times 96.0 = 8.66\,g$ ✓

(iv) The crucial point about a buffer is that the ratio [acid]:[salt] must not alter significantly when small amounts of acid or alkali are added. This can only happen if the acid and salt are present in similar amounts. Propanoic acid is a weak acid and so a solution has a significant $[CH_3CH_2COOH]$ but a tiny $[CH_3CH_2COO^-]$ ✓.

When OH^- ions are added they react with propanoic acid molecules:
$OH^- + C_2H_5COOH \rightarrow C_2H_5COO^- + H_2O$ ✓.

This causes $[CH_3CH_2COO^-]$ to rise significantly, which alters the [acid]:[salt] ratio and affects the pH ✓.

Questions & Answers

Question 4

(a) (i) Which of the following does **not** have a positive value of ΔS_{system}?

    A $MgSO_4(s) + aq \rightarrow Mg^{2+}(aq) + SO_4^{2-}(aq)$

    B $Ba(OH)_2.8H_2O(s) + 2NH_4Cl(s) \rightarrow BaCl_2(s) + 10H_2O(l) + 2NH_3(g)$

    C $NaCl(s) \rightarrow NaCl(l)$

    D $C(s) + H_2O(g) \rightarrow CO(g) + H_2(g)$ (1 mark)

(ii) An exothermic reaction is thermodynamically spontaneous at 20°C. Which of the following statements is correct?   (1 mark)

    A It will be spontaneous at all temperatures whatever the sign of ΔS_{system}.

    B It will only proceed at 20°C if the activation energy is not too high.

    C It will only proceed if ΔS_{system} is positive.

    D The value of ΔS_{total} is larger at 40°C than at 20°C.

(b) Consider the reaction: $CaSO_4(s) \rightarrow CaO(s) + SO_3(g)$

Substance	$CaSO_4(s)$	$CaO(s)$	$SO_3(g)$
Entropy / $J K^{-1} mol^{-1}$	107	40	256
$\Delta_f H$ / $kJ mol^{-1}$	−1434	−635	−396

(i) Calculate $\Delta S^{\ominus}_{system}$.   (2 marks)

ⓔ What are the units of S and hence of ΔS_{system}?

(ii) Calculate ΔH for the reaction and hence ΔS_{surr} at 500°C.   (3 marks)

ⓔ What are the units of ΔH and of T and, hence, of ΔS_{surr}?

(iii) Use your answers to (i) and (ii) to calculate ΔS_{total} at 500°C and hence the value of the equilibrium constant, K, at this temperature. $(R = 8.31 J K^{-1} mol^{-1})$ (3 marks)

ⓔ You must state the relationship between ΔS_{total} and K and then evaluate K.

(iv) Calculate the value ΔS_{surr} at 1000°C and hence comment on why the equilibrium constant is larger at this temperature than at 500°C. You may assume that neither ΔH nor ΔS_{system} alters.   (3 marks)

                              Total: 13 marks

Student answers

(a) (i) A ✓

ⓔ Even though there is a large increase in entropy of the solute going from solid to aqueous solution, the water is ordered by the high charge density of the group 2 cation. This more than outweighs the increase in entropy of the magnesium sulfate. Therefore ΔS_{system} is negative, making option A the correct answer to this negative question. It is worth putting '+' or '−' beside each response. Then the one marked '−' will be your answer. In B, a highly disordered gas is produced, so ΔS_{system} is positive. A molten liquid always has a higher entropy than its solid, so option C is positive. In D, 2 moles of gas are formed from 1 mole of gas, which causes an increase in entropy. Therefore, D is positive. The only answer with a negative ΔS_{system} is A.

(ii) B ✓

ⓔ For a thermodynamically feasible reaction to be observed, the activation energy must be reasonably low, so that the reactants are kinetically unstable as well as being thermodynamically unstable. Thus B is the correct response. It is worth checking the others. For a reaction to be spontaneous, ΔS_{total} must be positive. Since the reaction is exothermic, ΔS_{surr} is positive. If ΔS_{system} is negative, the reaction will cease to be spontaneous as the temperature is increased, because ΔS_{surr} will become less positive. This means that A is false. ΔS_{system} need not be positive. If it is negative and numerically less than the positive ΔS_{surr}, the reaction is feasible. Thus option C is false. ΔS_{total} is not larger at 40°C. It is an exothermic reaction and so the value of ΔS_{total} (and hence the equilibrium constant K) will decrease with an increase in temperature. Thus option D is also false.

(b) (i) $\Delta S_{system} = S(CaO) + S(SO_3) - S(CaSO_4)$ ✓
    $= +40 + 256 - (+107) = +189\,J\,K^{-1}\,mol^{-1}$ ✓

(ii) $\Delta H = \Delta_f H\,CaO + \Delta_f H\,SO_3 - \Delta_f H\,CaSO_4 = (-635) + (-396) - (-1434) = +403\,kJ\,mol^{-1}$ ✓
    $\Delta S_{surr} = -\Delta H/T$ ✓ $= -(+403\,000)\,J\,mol^{-1}/773\,K = -521\,J\,K^{-1}\,mol^{-1}$ ✓

ⓔ Be careful with units.

(iii) $\Delta S_{total} = \Delta S_{system} + \Delta S_{surroundings}$
    $= +189 + (-521) = -332\,J\,K^{-1}\,mol^{-1}$ ✓
    $\Delta S_{total} = R\ln K$
    $\ln K = \dfrac{\Delta S_{total}}{R} = \dfrac{-332}{8.31} = -39.95$ ✓
    $K = e^{-39.95} = 4.5 \times 10^{-18}$ ✓

(iv) $\Delta S_{surr} = -\dfrac{(+403\,000)}{1273} = -317\,J\,K^{-1}\,mol^{-1}$ ✓
    This is less negative than its value at 773 K ✓.
    Therefore, ΔS_{total} and hence K increase ✓.

Question 5

(a) ■ The lattice energy of calcium sulfate, $CaSO_4$, is $-2677\,kJ\,mol^{-1}$.
■ The hydration enthalpy of the $Ca^{2+}(g)$ ion is $-1650\,kJ\,mol^{-1}$; the hydration enthalpy of the $SO_4^{2-}(g)$ ion is $-1045\,kJ\,mol^{-1}$.
Draw a Hess's law diagram for the dissolving of calcium sulfate in water and calculate its enthalpy of solution. (4 marks)

ⓔ Make sure that the arrows on the Hess's law diagram go in the direction that agrees with the label.

(b) Which of strontium sulfate and barium sulfate is more soluble? Justify your answer using the data below to calculate the difference in ΔS_{total}. You may assume that the values are at a temperature of 298 K.

- $\Delta_{sol}H^{\ominus}$ $SrSO_4(s)$ = $-9\,kJ\,mol^{-1}$
- $\Delta_{sol}H^{\ominus}$ $BaSO_4(s)$ = $+19\,kJ\,mol^{-1}$
- $S^{\ominus}$ $Sr^{2+}(aq)$ = $-33\,J\,K^{-1}\,mol^{-1}$
- $S^{\ominus}$ $Ba^{2+}(aq)$ = $+10\,J\,K^{-1}\,mol^{-1}$

(5 marks)

ⓔ You need to work out the *difference* between ΔS_{surr} of the two salts dissolving and the *difference* of the two ΔS_{system} values. Use these to evaluate which has the more positive ΔS_{total}. Note that the entropy of $SO_4^{2-}(aq)$ is not required, as it common to both compounds.

Total: 9 marks

Student answers

(a)

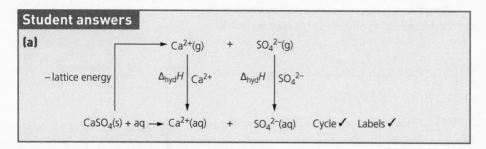

Cycle ✓   Labels ✓

ⓔ The arrow can be drawn from $Ca^{2+}(g)$ + $SO_4^{2-}(g)$ to $CaSO_4(s)$ in which case it must be labelled lattice energy (not – lattice energy).

$\Delta H_{solution}$ = – lattice energy + sum of hydration energies of ions ✓.

= $-(-2677) + (-1650) + (-1045) = -18\,kJ\,mol^{-1}$ ✓.

(b) First and second marks:

either ΔS_{surr} for $BaSO_4$ is $-\dfrac{(+19000)}{298} = -64\,J\,K^{-1}\,mol^{-1}$ and for $SrSO_4 = -\dfrac{(-9000)}{298}$

= $+30\,J\,K^{-1}\,mol^{-1}$ ✓.

So ΔS_{surr} for barium sulfate is $94\,J\,K^{-1}\,mol^{-1}$ less positive ✓.

or

The value of $\Delta H_{solution}$ for barium sulfate is more endothermic by $28\,kJ\,mol^{-1}$ ✓.

This means that ΔS_{surr} for barium sulfate becomes $\dfrac{28000}{298} = 94\,J\,K^{-1}\,mol^{-1}$

more negative ✓.

Remaining three marks:

The value of ΔS_{system} of barium sulfate is $43\,J\,K^{-1}\,mol^{-1}$ more positive than that for strontium sulfate ✓.

As $\Delta S_{total} = \Delta S_{system} + \Delta S_{surr}$, the value of ΔS_{system} for barium sulfate is $94 - 43 = 51\,J\,K^{-1}\,mol^{-1}$ less positive than that of strontium sulfate ✓, so barium sulfate will be less soluble ✓ than strontium sulfate.

(e) The answer can be argued comparing strontium sulfate to barium sulfate, ending with the statement that strontium sulfate will be more soluble than barium sulfate.

Question 6

This question is about redox potentials.

(a) (i) When the standard electrode potential of zinc is measured, a reference electrode is needed because: (1 mark)

    A the potential between a metal and its solution cannot be measured directly

    B a standard hydrogen electrode has a potential of zero

    C zinc is the cathode

    D a pressure of 100 kPa must be used

(ii) If the standard electrode potential of a cell, $E^{\ominus}_{cell}$, is +0.10 V, the reaction: (1 mark)

    A is not thermodynamically feasible, as the value of $E^{\ominus}_{cell}$ is too small

    B will always take place

    C will take place under standard conditions, if the activation energy is not too high

    D will have a negative value of ΔS_{total}

(b) Use items 8 and 28 in the Edexcel *Data Booklet* and the value below to answer the following questions.

$$S(s) + 2H^+(aq) + 2e^- \rightleftharpoons H_2S(aq) \qquad E^{\ominus} = 0.14\,V$$

(e) Remember that all redox half-equations are written as reductions with the electrons on the left. One of the reactants will be on the right of one of the half-equations.

(i) Hydrogen sulfide (H_2S) reduces dichromate(vi) ions in acid solution. Is the product Cr^{3+} ions or Cr^{2+} ions? Justify your answer, calculating the $E^{\ominus}_{cell}$ values for possible reductions. (4 marks)

(ii) Write the ionic equation for the reaction that takes place between hydrogen sulfide and dichromate(vi) ions in acid solution. (2 marks)

(iii) Define disproportionation. (1 mark)

(iv) Explain, in terms of E values, whether $Cr^{3+}(aq)$ ions will disproportionate. (2 marks)

(e) Disproportionation is when an element in a single species undergoes simultaneous oxidation and reduction.

(c) The water from volcanic hot springs often contains significant amounts of the poisonous gas hydrogen sulfide. The amount present can be measured by titration against a standard solution of acidified potassium manganate(vii).

$$5H_2S + 2MnO_4^- + 6H^+ \rightarrow 2Mn^{2+} + 5S + 8H_2O$$

A portion of hot spring water of volume 250 cm^3 was taken and 50.0 cm^3 portions were titrated against acidified 0.0100 mol dm^{-3} potassium manganate(vii) solution. The mean titre was 7.55 cm^3. Calculate the volume of hydrogen sulfide dissolved in the 250 cm^3 sample of hot spring water. (1 mol of gas, under the conditions of the experiment, has a volume of 24 000 cm^3.) (4 marks)

Total: 15 marks

Questions & Answers

(a) (i) A ✓

e Zinc ions go into solution and the electrons remain on the zinc rod, thus setting up a potential difference. However, this cannot be measured directly as any connection to a voltmeter using a metal wire dipping into the solution would alter the potential. A salt bridge has to be used, which is then connected to a reference electrode. Options B and D are true statements but neither is the reason why a reference electrode has to be used. Option C is incorrect, as the zinc becomes oxidised and is, therefore, the anode.

(ii) C ✓

e It is a common misunderstanding that the cell potential has to be above a value, such as $>+0.03\,V$ or $>+0.3\,V$. As long as the cell potential is positive, the reaction is thermodynamically feasible. Therefore, option A is incorrect. Option B is incorrect because the reaction may not happen for kinetic reasons. Option D is incorrect because a positive value for the cell potential means that ΔS_{total} is also positive.

(b) (i) Either:

$$Cr^{3+}(aq) + e^- \rightleftharpoons Cr^{2+}(aq) \qquad\qquad E = -0.41\,V \quad \text{equation 1}$$

$$S(s) + 2H^+(aq) + 2e^- \rightleftharpoons H_2S(aq) \qquad\qquad E = +0.14\,V \quad \text{equation 2}$$

$$Cr_2O_7^{2-}(aq) + 14H^+(aq) + 6e^- \rightleftharpoons 2Cr^{3+}(aq) + 7H_2O(l) \quad E = +1.33\,V \quad \text{equation 3}$$

✓ for all three.
Using the anticlockwise rule on equations 2 and 3, $E^{\ominus}_{cell}$ is $(+1.33) + (-0.14) = +1.19\,V$ ✓. This is positive so the reduction of $Cr_2O_7^{2-}$ to Cr^{3+} ions will happen ✓.
Using the rule on equations 1 and 2, $E^{\ominus}_{cell}$ is $-0.41 + (-0.14) = -0.55\,V$. This is negative so the reduction will not happen, and so H_2S will reduce chromate(vi) ions to Cr^{3+} and no further ✓.

e Using the anticlockwise rule for processes that are thermodynamically not feasible is difficult. The rule on equations 1 and 2 would mean that sulfur would be reduced to H_2S by Cr^{2+} ions, so obviously the reverse would have a negative $E^{\ominus}$cell value.

Or:

$$Cr^{3+}(aq) + e^- \rightleftharpoons Cr^{2+}(aq) \qquad E = -0.41\,V$$
$$Cr_2O_7^{2-}(aq) + 14H^+(aq) + 6e^- \rightleftharpoons 2Cr^{3+}(aq) + 7H_2O(l) \qquad E = +1.33\,V$$
$$S(s) + 2H^+(aq) + 2e^- \rightleftharpoons H_2S(aq) \qquad E = +0.14\,V$$

✓ for all three.

Reduction to Cr^{3+}:

The third equation has to be reversed.

The value of $E^\ominus_{cell}$ is $(+1.33) + (-0.14) = +1.19\,V$ ✓

This is positive, so hydrogen sulfide will reduce chromium(VI) ions to chromium(III) ions ✓.

ⓔ Remember that when a redox half-equation is reversed, the sign of E must be altered; when it is multiplied by a number, the E value is unaltered.

Reduction to Cr^{2+}:

The first equation has to be multiplied by 2 and the third equation reversed:

$$2Cr^{3+}(aq) + 2e^- \rightleftharpoons 2Cr^{2+}(aq) \qquad E^\ominus = -0.41\,V$$
$$H_2S(aq) \rightleftharpoons 2H^+(aq) + S(s) + 2e^- \qquad E^\ominus = -(+0.14\,V) = -0.14\,V$$

The value of $E^\ominus_{cell}$ is $-0.41 + (-0.14) = -0.55\,V$ ✓. This is negative so hydrogen sulfide will reduce chromium(VI) ions to chromium(III) ions but no further ✓.

or:

For the reduction to Cr^{3+}:

$E^\ominus_{cell} = (E^\ominus$ of the oxidising agent$) - (E^\ominus$ of the reducing agent$)$ ✓ $= +1.33 - (+0.14) = +1.19\,V$ ✓. This is positive, so the reaction will happen ✓.

For the reduction from Cr^{3+} to Cr^{2+}:

$E^\ominus_{cell} = (E^\ominus$ of the oxidising agent$) - (E^\ominus$ of the reducing agent$) = -0.41 + (-0.14) = -0.55\,V$ This is negative, so the reaction will not happen ✓.

(ii) The overall equation is obtained by adding:

$$Cr_2O_7^{2-}(aq) + 14H^+(aq) + 6e^- \rightarrow 2Cr^{3+}(aq) + 7H_2O(l)$$
$$3H_2S(aq) \rightarrow 6H^+(aq) + 3S(s) + 6e^- ✓$$
$$Cr_2O_7^{2-}(aq) + 8H^+(aq) + 3H_2S \rightarrow 2Cr^{3+}(aq) + 7H_2O(l) + 3S(s) ✓$$

ⓔ You must remember to cancel the electrons and the H^+ ions: $14H^+$ on the left-hand side and $6H^+$ on the right-hand side cancel to leave $8H^+$ on the left.

(iii) Disproportionation is when an element in a single species is simultaneously oxidised and reduced ✓.

(iv) The disproportionation reaction would involve Cr^{3+} ions being reduced to Cr^{2+} ions (item 8 in the *Data Booklet*) and Cr^{3+} ions being oxidised to $Cr_2O_7^{2-}$ ions (item 28 reversed) ✓.

The $E^{\ominus}_{cell}$ for this is (E for the first half-equation) − (E for the second half-equation):

$E^{\ominus}_{cell} = (-0.41) - (+1.33) = -1.74\,V$. This value is negative and so disproportionation of Cr^{3+} ions does not occur ✓.

(c) amount (moles) of $MnO_4^- =$ concentration × volume

$= 0.0100\,mol\,dm^{-3} \times 0.00755\,dm^3$

$= 7.55 \times 10^{-5}\,mol$ ✓

amount (moles) of H_2S in $50\,cm^3 = \dfrac{5}{2} \times 7.55 \times 10^{-5} = 1.8875 \times 10^{-4}\,mol$ ✓

amount (moles) in $250\,cm^3 = \dfrac{250}{50} \times 1.8875 \times 10^{-4} = 9.4375 \times 10^{-4}\,mol$ ✓

volume of hydrogen sulfide dissolved $= 9.4375 \times 10^{-4} \times 24\,000 = 22.7\,cm^3$ ✓

ⓔ You must state at each stage what you are calculating so that consequential marks can be awarded if an error is made in an earlier step. A jumble of numbers would not allow this.

Question 7

(a) Which of the following is the electron configuration of a stable, coloured transition-metal ion?   (1 mark)

    A [Ar] $3d^{10}\,4s^1$

    B [Ar] $3d^{10}\,4s^0$

    C [Ar] $3d^6\,4s^1$

    D [Ar] $3d^6\,4s^0$

(b) (i) Explain, with the aid of an equation, why hydrated chromium(III) ions in solution are acidic.   (2 marks)

ⓔ The reaction is between the chromium ions and water. What ions make a solution acidic?

    (ii) State what you would see when aqueous sodium hydroxide is added steadily until in excess to a solution of hydrated chromium(III) ions. Give the equations for the reactions.   (4 marks)

ⓔ The question asks for observations and equations in the plural, so there are two reactions, the second being with excess alkali.

(c) (i) EDTA reacts with hydrated chromium(III) ions to form a polydentate complex ion in an exothermic reaction:

    $EDTA^{4-}(aq) + [Cr(H_2O)_6]^{3+}(aq) \rightarrow [Cr(EDTA)]^-(aq) + 6H_2O(l)$

    Explain why ΔS_{total} is positive for this reaction.   (2 marks)

ⓔ You must make a comment about ΔS_{system} and ΔS_{surr}.

(ii) The EDTA complex ion is coloured. Explain why it is coloured. (3 marks)

(iii) When blue solid hydrated copper sulfate ($CuSO_4.5H_2O$) is heated, it produces steam and white anhydrous copper sulfate ($CuSO_4$). This compound is white because: (1 mark)

A copper has an electron configuration of [Ar] $3d^{10} 4s^1$

B the copper ion in $CuSO_4$ has a full d-shell

C the copper ion no longer has any ligands and so the d-orbitals are not split

D the copper ions are reduced to Cu^+ ions

Total: 9 marks

ⓔ You must explain what effect the ligand has on the electrons in the Cr^{3+} ion and what happens when light is absorbed.

Student answers

(a) D ✓.

ⓔ This is the electron configuration of an Fe^{2+} ion. Option A is the configuration of a copper atom — note that the question asks about a transition metal *ion*. Option B is the configuration of Zn^{2+} or Cu^+. Zinc is not a transition metal, as its ion does not have an unpaired d-electron and copper(ı) ions are not coloured. Option C is the configuration for an Fe^+ ion, which does not exist.

(b) (i) The hydrated ions are reversibly deprotonated by water forming H_3O^+ ions, which make the solution acidic ✓.
$[Cr(H_2O)_6]^{3+}(aq) + H_2O(l) \rightleftharpoons [Cr(H_2O)_5OH]^{2+}(aq) + H_3O^+(aq)$ ✓

ⓔ The first mark is for the formation of H_3O^+ ions.

(ii) At first a dirty green precipitate ✓ is formed.
$[Cr(H_2O)_6]^{3+}(aq) + 3OH^-(aq) \rightarrow [Cr(H_2O)_3(OH)_3](s) + 3H_2O(l)$ ✓
This then dissolves in excess sodium hydroxide to form a green solution ✓.
$[Cr(H_2O)_3(OH)_3](s) + 3OH^-(aq) \rightarrow [Cr(OH)_6]^{3-}(aq) + 3H_2O(l)$ ✓

(c) (i) The reaction with EDTA involves two particles reacting to form seven particles, all in the aqueous phase. This is a significant increase in disorder and so ΔS_{system} is positive ✓. The question states that the reaction is exothermic, so ΔS_{surr} will be positive as well because $\Delta S_{surr} = -\Delta H/T$. As both entropy terms are positive, ΔS_{total} will be positive ✓.

(ii) The six pairs of electrons forming the bonds between EDTA and the Cr^{3+} ion cause the d-orbitals in the chromium ion to split into two levels ✓. When white light is shone onto the complex, some of the frequencies are absorbed ✓ and an electron is promoted from the lower level of the split d-orbitals to a higher level ✓. This causes the ion to have the complementary colour to that of the wavelength absorbed.

ⓔ Do not confuse why ions are coloured with the colour in a flame test.

> **(iii)** C ✓

e Solid hydrated copper sulfate has four water ligands. When it is heated, these are lost. As there are now no ligands, there is no splitting of the d-orbitals. Option A is a true statement about a copper atom, but is irrelevant to this question. Option B is incorrect as the copper ion in $CuSO_4$ is 2+ and has the electron configuration $[Ar]\ 3d^9$. Option D is incorrect because hydrated copper(II) sulfate is not reduced on heating.

Question 8

(a) Measuring $E^{\ominus}$ for a half-cell involves having the species at the correct concentrations and using a reference electrode.

e Think carefully about the necessary concentrations and other standard conditions.

(i) A standard hydrogen electrode consists of: (1 mark)

	Electrode	Electrolyte	Temperature/°C
A	Platinum	½ mol dm^{-3} H$_2$SO$_4$	273
B	Platinum	1 mol dm^{-3} HCl	298
C	Mercury	1 mol dm^{-3} HNO$_3$	273
D	Mercury	Saturated Hg$_2$Cl$_2$ + KCl	298

(ii) When measuring $E^{\ominus}$ for $MnO_4^-(aq) + 8H^+(aq) + 5e^- \rightleftharpoons Mn^{2+}(aq) + 4H_2O(l)$
the concentrations must be: (1 mark)

	MnO$_4^-$(aq)	H$^+$(aq)	Mn^{2+}(aq)
A	1.0 mol dm^{-3}	1.0 mol dm^{-3}	1.0 mol dm^{-3}
B	1.0 mol dm^{-3}	8.0 mol dm^{-3}	1.0 mol dm^{-3}
C	0.10 mol dm^{-3}	0.10 mol dm^{-3}	0.10 mol dm^{-3}
D	0.10 mol dm^{-3}	0.80 mol dm^{-3}	0.10 mol dm^{-3}

(b) (i) Construct the cell diagram, using conventional representations of half-cells with platinum electrodes, for the reduction of $IO_3^-(aq)$ ions in acid solution by $I^-(aq)$ ions.
$IO_3^-(aq) + 6H^+(aq) + 5I^-(aq) \rightarrow 3I_2(aq) + 3H_2O(l)$ (4 marks)

(ii) The concentration of a solution of potassium iodate(v), KIO_3, was found as follows. 25.0 cm^3 portions of the potassium iodate(v) solution were taken and added to excess dilute sulfuric acid and potassium iodide solutions. They were titrated against a solution of sodium thiosulfate of concentration 0.100 mol dm^{-3} using starch as an indicator. The mean titre was 24.35 cm^3.
$I_2(aq) + 2S_2O_3^{2-}(aq) \rightarrow 2I^-(aq) + S_4O_6^{2-}(aq)$
Calculate the concentration of the potassium iodate(v) solution in mol dm^{-3}. (4 marks)

e Use the stoichiometry to work out the moles of iodine reacted with the sodium thiosulfate and hence the number of moles of iodate(v) ions.

(iii) If the starch were added too soon, an inaccurate titre would be obtained. Explain why it would be inaccurate.

(1 mark)

Total: 11 marks

Student answers

(a) (i) B ✓

e Sulfuric acid is only fully ionised in its 1st ionisation, so $[H^+]$ will be less than $1.0\,mol\,dm^{-3}$. Also the temperature should be 298 K not 273 K. Options C and D are a confusion with a calomel electrode.

(ii) A ✓

e All concentrations must be $1\,mol\,dm^{-3}$ regardless of the stoichiometry.

(b) (i) $Pt(s) \mid 5I^-(aq), 2\frac{1}{2}I_2(aq), \parallel IO_3^-(aq), 6H^+(aq), \frac{1}{2}I_2(aq) \mid Pt(s)$

Pt electrode ✓, left-hand half cell ✓, right-hand half cell ✓ all state symbols correct ✓.

e The order is: electrode then left cell (reducing agent + its oxidised product) then the right cell (oxidising agent + its reduced product), then the other electrode.

(ii) moles thiosulfate = $0.02435\,dm^3 \times 0.100\,mol\,dm^{-3} = 0.002435\,mol$ ✓

moles I_2 liberated = $\frac{1}{2} \times 0.002435 = 0.0.0012175\,mol$ ✓

moles $KIO_3 = \frac{1}{3} \times 0.0.0012175 = 4.058 \times 10^{-4}\,mol$ ✓

concentration $KIO_3 = \dfrac{4.058 \times 10^{-4}\,mol}{0.0250\,dm^3} = 0.0162\,mol\,dm^{-3}$ ✓

e Be careful about the stoichiometry between thiosulfate and iodine, and between iodine and iodate.

(iii) The starch would form an insoluble blue-black precipitate with the iodine ✓.

Knowledge check answers

1 a As there are the same numbers of gas moles on each side of the equation, the units cancel.

b $K_c = \dfrac{[HI]^2}{[H_2][I_2]} = 49$

$[HI]^2 = K_c \times [H_2][I_2]$

$[H_2] = [I_2] = 0.0040\,mol\,dm^{-3}$

$[HI] = \sqrt{49 \times 0.0040 \times 0.0040} = 0.028\,mol\,dm^{-3}$

2 $p(N_2) + p(H_2) + p(NH_3) = P(total)$

$p(NH_3) = P(total) - p(N_2) - p(H_2) = 100 - 19 - 57 = 24\,atm$

$K_p = \dfrac{p(NH_3)^2}{p(N_2)\,p(H_2)^3} = \dfrac{24^2}{19 \times 57^3} = 1.6 \times 10^{-4}\,atm^{-2}$

3 First ionisation: $H_2SO_4 + H_2O \rightarrow H_3O^+ + HSO_4^-$

Second ionisation: $HSO_4^- + H_2O \rightleftharpoons H_3O^+ + SO_4^{2-}$

4 $HOI + CH_3COOH \rightarrow H_2OI^+ + CH_3COO^-$

5 $pOH = -\log(0.0246) = 1.61$

6 pH = 2.97, so $[H^+] = 10^{-2.97} = 0.0010715\,mol\,dm^{-3}$
$= [C_2H_5COO^-]$

$K_a = \dfrac{[H^+][C_2H_5COO^-]}{[C_2H_5COOH]}$

$[C_2H_5COOH] = \dfrac{[H^+][C_2H_5COO^-]}{K_a} = \dfrac{(0.0010715)^2}{1.35 \times 10^{-5}}$

$= 0.0850\,mol\,dm^{-3}$

Note that if $[H^+]$ had been used at 3 significant figures, the final answer would have been $0.0848\,mol\,dm^{-3}$.

7 a Red to orange (note that the indicator would go yellow, if the end point had been overshot).

b Colourless to very pale pink.

8 The added OH^- ions can be removed by the reaction:
$HA + OH^- \rightarrow A^- + H_2O$

The value of [HA] will decrease by an insignificant amount but the value of [A⁻] will increase significantly, as it was very low because there is no salt present. This causes the ratio [HA] : [A⁻] to change significantly and so the pH will also change (rise) significantly.

9 $NaCl < MgCl_2 < MgBr_2$

10 $K^+ < Na^+ < Ca^{2+} < Mg^{2+}$

11 a Cr is +6 (7 × −2 = −14, but ion is 2−, so 2 × Cr = +12 or +6 each)

b I is +5 (3 × −2 = −6, but the ion is 1−, so I = +5)

12 The oxidation number of manganese in MnO_4^- is +7 and in Mn^{2+} it is +2. That of carbon in $C_2O_4^{2-}$ is +3 each or +6 in total, and in CO_2 it is +4. The oxidation number of manganese decreases by 5 and that of each of the two carbon atoms increases by 1, so there must be two MnO_4^- ions and five $C_2O_4^{2-}$ ions so that the total change of each is 10. In order to balance charge and oxygen atoms, there must be $16H^+$ on the left and $8H_2O$ on the right. Thus the equation is:

$2MnO_4^- + 5C_2O_4^{2-} + 16H^+ \rightarrow 2Mn^{2+} + 10CO_2 + 8H_2O$

13 The electrode consists of a platinum plate dipping into a solution that is $1\,mol\,dm^{-3}$ in MnO_4^- ions, H^+ ions and Mn^{2+} ions, all at a temperature of 25°C.

14 $E^\ominus_{cell} = +1.19 - (+1.03) = +0.16\,V$, so iodate(V) ions will oxidise nitrogen(II) oxide because $E^\ominus_{cell}$ is positive.

15 Either:

$\frac{1}{2}Cl_2 + e^- \rightleftharpoons Cl^- \qquad E^\ominus = +1.36\,V$

$HClO + H^+ + e^- \rightleftharpoons \frac{1}{2}Cl_2 + H_2O \qquad E^\ominus = +1.59\,V$

Reversing the first equation and adding it to the second gives the overall equation.

$Cl^- \rightleftharpoons \frac{1}{2}Cl_2 + e^- \qquad E^\ominus = -(+1.36) = -1.36\,V$

$HClO + H^+ + e^- \rightleftharpoons \frac{1}{2}Cl_2 + H_2O \qquad E^\ominus = +1.59\,V$

$HClO + H^+ + Cl^- \rightarrow Cl_2 + H_2O$
$\qquad\qquad E^\ominus_{cell} = +1.59 + (-1.36) = +0.23\,V$

The value is positive, so the reaction is feasible.

Or:

$Cl_2(aq), 2Cl^-(aq) \qquad E^\ominus = +1.36\,V$

$[2HClO(aq) + 2H^+(aq)], [Cl_2(aq) + 2H_2O(l)] \quad E^\ominus = +1.59\,V$

Using the anticlockwise rule, HClO, in acid solution (on the left of the lower data), will oxidise Cl^- ions (on the right of the upper data) to Cl_2 with $E^\ominus_{cell} = +1.59 - (+1.36) = +0.23\,V$.

The value is positive, so the reaction is feasible.

As the change in oxidation number is 1 per chlorine atom in both, you need 2HClO for each Cl_2 so that both half-equations will have $2e^-$, so the overall equation is:

$2HClO(aq) + 2H^+(aq) + 2Cl^-(aq) \rightarrow 2Cl_2(aq) + 2H_2O$

(This can now be divided by 2.)

16 $3MnO_4^{2-} + 4H^+ \rightarrow 2MnO_4^- + MnO_2 + 2H_2O$

Note that the change in oxidation number going from MnO_4^{2-} to MnO_4^- is +1 and that from MnO_4^{2-} to MnO_2 it is −2, so there must be three MnO_4^{2-} ions, two of which are oxidised to MnO_4^- and one reduced to MnO_2.

17 amount of thiosulfate in titre = $0.100\,mol\,dm^{-3} \times 0.02645\,dm^3 = 0.002645\,mol$

amount of iodine released = $\frac{1}{2} \times 0.002645$

$= 0.0013225\,mol$

amount of IO_3^- ions in $25\,cm^3 = \frac{1}{3} \times 0.0013225$

$= 0.0004408\,mol$

amount of IO_3^- ions in $250\,cm^3 = 10 \times 0.0004408$
$= 0.004408\,mol$

molar mass of $MIO_3 = \dfrac{mass}{moles} = \dfrac{0.87\,g}{0.004408\,mol} = 197\,g\,mol^{-1}$

relative atomic mass of M = 197 − (127 + (16 × 3)) = 22

The nearest group 1 metal is sodium, so M is sodium.

18 amount of MnO_4^- in titre = $0.0100\,mol\,dm^{-3} \times 0.0168\,dm^3$
$= 0.000168\,mol$

amount of $H_2C_2O_4$ in sample = $\frac{5}{2} \times 0.000168$

$= 0.000420\,mol$

mass $H_2C_2O_4$ in sample = $0.000420\,mol \times 90\,g\,mol^{-1}$
$= 0.0378\,g$

% of ethanedioic acid in rhubarb leaves $= \dfrac{0.0378 \times 100}{3.14}$

$= 1.20\%$

19 a Cr^{3+} is [Ar], $3d^3$

b Mn^{2+} is [Ar], $3d^5$ (you can add $4s^0$ to both if you wish)

20 The ring that would have to form between the two nitrogen atoms and the central metal ion would only contain four atoms. The resulting strain would make its formation energetically impossible.

21 a As there are no ligands in anhydrous copper(II) sulfate, the *d*-orbitals are not split. This means that light cannot be absorbed.

b The electronic configuration of copper(I) ions is [Ar], $3d^{10}$. Even though the five *d*-orbitals are split in energy, all five are full and so no *d–d* transitions can take place.

22 *Either:*

The $E^\ominus$ value given is less positive than that for VO_2^+ to VO^{2+} and for VO^{2+} to V^{3+} but more positive than that for V^{3+} to V^{2+}, so H_2S will reduce VO_2^+ to VO^{2+} and then to V^{3+} but no further.

Or: Using the half-equation and reversing it:

$H_2S \rightleftharpoons S + 2H^+ + 2e^-$ $E^\ominus = -(+0.14) = -0.14\,V$

Adding this to the $5 \rightarrow 4$ half-equation in Table 16 on page 58 which has $E^\ominus = 1.00\,V$, gives a positive $E^\ominus_{cell}$ and so reduction occurs.

Likewise adding it to the $4 \rightarrow 3$ half-equation also gives a positive $E^\ominus_{cell}$, so reduction will go to V^{3+}. However, adding it to the $3 \rightarrow 2$ half-equation gives a negative $E^\ominus_{cell}$ and so H_2S will reduce VO_2^+ ions to V^{3+} but no further.

23 The added OH^- ions will remove H^+ from the equilibrium reaction, thus driving it to the left. Therefore the colour will change from orange to yellow.

24 $E^\ominus_{cell} = E^\ominus$(oxidising agent) – $E^\ominus$(reducing agent)

For +6 to +3: $E^\ominus_{cell} = E^\ominus(Cr_2O_7^{2-}/Cr^{3+}) - E^\ominus(Br_2/Br^-)$ $=1.33 - (+1.07) = +0.26\,V$, so it will reduce it to Cr^{3+}.

For +3 to +2: $E^\ominus_{cell} = E^\ominus(Cr^{3+}/Cr^{2+}) - E^\ominus(Br_2/Br^-) =$ $-0.41 - (+1.07) = -1.48\,V$, so it will not reduce Cr^{3+}.

25 Using method 1 on page 43:

$E^\ominus_{cell} = E^\ominus$(oxidising agent) – $E^\ominus$(reducing agent) = $E^\ominus(Fe^{3+}/Fe^{2+}) - E^\ominus(Cr_2O_7^{2-}/Cr^{3+}) = +0.77 - (1.33) =$ $-0.56\,V$. This is a negative number, so Fe^{3+} ions will not oxidise Cr^{3+} ions in acid solution.

Or using the anticlockwise rule (more difficult this time):

$Fe^{3+} + e^- \rightleftharpoons Fe^{2+}$ $E^\ominus = +0.77\,V$

$Cr_2O_7^{2-} + 14H^+ + 6e^- \rightleftharpoons 2Cr^{3+} + 7H_2O$ $E^\ominus = +1.33\,V$

Going anticlockwise from bottom left: $Cr_2O_7^{2-}$ ions would react with Fe^{2+} ions with $E^\ominus_{cell} = +0.56\,V$, so the *opposite* reaction of Cr^{3+} being oxidised by Fe^{3+} will not happen.

Index

The periodic table

Key:

Relative atomic mass
Atomic symbol
name
Atomic (proton) number

Group

Period	1	2												3	4	5	6	7	0
1	1.0 **H** hydrogen 1																		4.0 **He** helium 2
2	6.9 **Li** lithium 3	9.0 **Be** beryllium 4												10.8 **B** boron 5	12.0 **C** carbon 6	14.0 **N** nitrogen 7	16.0 **O** oxygen 8	19.0 **F** fluorine 9	20.2 **Ne** neon 10
3	23.0 **Na** sodium 11	24.3 **Mg** magnesium 12												27.0 **Al** aluminium 13	28.1 **Si** silicon 14	31.0 **P** phosphorus 15	32.1 **S** sulfur 16	35.5 **Cl** chlorine 17	39.9 **Ar** argon 18
4	39.1 **K** potassium 19	40.1 **Ca** calcium 20	45.0 **Sc** scandium 21	47.9 **Ti** titanium 22	50.9 **V** vanadium 23	52.0 **Cr** chromium 24	54.9 **Mn** manganese 25	55.8 **Fe** iron 26	58.9 **Co** cobalt 27	58.7 **Ni** nickel 28	63.5 **Cu** copper 29	65.4 **Zn** zinc 30		69.7 **Ga** gallium 31	72.6 **Ge** germanium 32	74.9 **As** arsenic 33	79.0 **Se** selenium 34	79.9 **Br** bromine 35	83.8 **Kr** krypton 36
5	85.5 **Rb** rubidium 37	87.6 **Sr** strontium 38	88.9 **Y** yttrium 39	91.2 **Zr** zirconium 40	92.9 **Nb** niobium 41	95.9 **Mo** molybdenum 42	[98] **Tc** technetium 43	101.1 **Ru** ruthenium 44	102.9 **Rh** rhodium 45	106.4 **Pd** palladium 46	107.9 **Ag** silver 47	112.4 **Cd** cadmium 48		114.8 **In** indium 49	118.7 **Sn** tin 50	121.8 **Sb** antimony 51	127.6 **Te** tellurium 52	126.9 **I** iodine 53	131.3 **Xe** xenon 54
6	132.9 **Cs** caesium 55	137.3 **Ba** barium 56	138.9 **La** lanthanum 57	178.5 **Hf** hafnium 72	180.9 **Ta** tantalum 73	183.8 **W** tungsten 74	186.2 **Re** rhenium 75	190.2 **Os** osmium 76	192.2 **Ir** iridium 77	195.1 **Pt** platinum 78	197.0 **Au** gold 79	200.6 **Hg** mercury 80		204.4 **Tl** thallium 81	207.2 **Pb** lead 82	209.0 **Bi** bismuth 83	[209] **Po** polonium 84	[210] **At** astatine 85	[222] **Rn** radon 86
7	[223] **Fr** francium 87	[226] **Ra** radium 88	[227] **Ac** actinium 89	[261] **Rf** rutherfordium 104	[262] **Db** dubnium 105	[266] **Sg** seaborgium 106	[264] **Bh** bohrium 107	[277] **Hs** hassium 108	[268] **Mt** meitnerium 109	[271] **Ds** darmstadtium 110	[272] **Rg** roentgenium 111								

Elements with atomic numbers 112–116 have been reported but not fully authenticated

140.1 **Ce** cerium 58	140.9 **Pr** praseodymium 59	144.2 **Nd** neodymium 60	144.9 **Pm** promethium 61	150.4 **Sm** samarium 62	152.0 **Eu** europium 63	157.2 **Gd** gadolinium 64	158.9 **Tb** terbium 65	162.5 **Dy** dysprosium 66	164.9 **Ho** holmium 67	167.3 **Er** erbium 68	168.9 **Tm** thulium 69	173.0 **Yb** ytterbium 70	175.0 **Lu** lutetium 71
232 **Th** thorium 90	[231] **Pa** protactinium 91	238.1 **U** uranium 92	[237] **Np** neptunium 93	[242] **Pu** plutonium 94	[243] **Am** americium 95	[247] **Cm** curium 96	[245] **Bk** berkelium 97	[251] **Cf** californium 98	[254] **Es** einsteinium 99	[253] **Fm** fermium 100	[256] **Md** mendelevium 101	[254] **No** nobelium 102	[257] **Lr** lawrencium 103